B

MATHEMATICAL MODELING

No. 2

Edited by
William F. Lucas, Claremont Graduate School
Maynard Thompson, Indiana University

Hua Loo-Keng   Wang Yuan

Popularizing Mathematical Methods in the People's Republic of China

Some Personal Experiences

Revised and Edited by J.G.C. Heijmans

Birkhäuser
Boston · Basel · Berlin

Hua Loo-Keng
(1911–1985)

Wang Yuan
Academia Sinica
Institute of Mathematics
Beijing
People's Republic of China

J.G.C. Heijmans
Department of Mathematics
University of Texas at Arlington
Arlington, TX 76019
U.S.A.

Library of Congress Cataloging-in-Publication Data
Hua, Loo-keng, 1910–
    Popularizing mathematical methods in the People's Republic of China:
Some personal experiences / by Hua Loo-keng and Wang Yuan ;
revised and edited by J.G.C. Heijmans.
        p. cm.—(Mathematical modeling ; 2)
    Includes bibliographies.
    ISBN 0-8176-3372-3 (alk. paper)
    1. Mathematical optimization.   2. Industrial engineering–
Mathematics.   3. Operations research.   I. Wang, Yüan, fl. 1963–
    II. Heijmans, J.G.C.   III. Title.   IV. Series: Mathematical
modeling ; no. 2.
QA402.5.H79 1989
519—dc19   88-39804
                                                            CIP

UNIVERSITY
LIBRARY
NOTTINGHAM

Printed on acid-free paper.

© Birkhäuser Boston, 1989

All rights reserved. No part of this publication may be reproduced, stored in a retrieval system, or transmitted, in any form or by any means, electronic, mechanical, photocopying, recording or otherwise, without prior permission of the copyright owner.

ISBN 0-8176-3372-3
ISBN 3-7643-3372-3

Text provided in camera-ready form by the editor.
Printed and bound by R.R. Donnelley & Sons, Harrisonburg, Virginia.

9 8 7 6 5 4 3 2 1

Contents

CHAPTER 0

INTRODUCTION

CHAPTER 1

ON THE CALCULATION OF MINERAL RESERVES AND HILLSIDE AREAS ON CONTOUR MAPS

CHAPTER 5

THE GOLDEN NUMBER AND NUMERICAL INTEGRATION

CHAPTER 6

OVERALL PLANNING METHODS

CHAPTER 7

PROGRAM EVALUATION AND REVIEW TECHNIQUE (PERT)

CHAPTER 8
MACHINE SCHEDULING

CHAPTER 9
THE TRANSPORTATION PROBLEM (GRAPHICAL METHOD)

CHAPTER 10
THE TRANSPORTATION PROBLEM (SIMPLEX METHOD)

CHAPTER 11
THE POSTMAN PROBLEM

Acknowledgements

I would like to express my gratitude first of all to Dr. H.O. Pollak of Bell Laboratories in the United States, for so kindly inviting me to present my report entitled "Some Personal Experiences in Popularizing Mathematical Methods in the People's Republic of China", both at the Fourth International Conference on Mathematical Education and at Bell Laboratories[1]. Dr. Pollak wrote to me on several occasions urging me to make additions to my report and collate the material for publication as a book with Birkhäuser Boston, Inc. The fact that this has now been done is due to his kind recommendation and sincere concern.

My special thanks go to all colleagues, workers and professionals who have been engaged in the popularization of mathematical methods over the past 20 years and have done a great deal of creative work in this field. Because of the vast number of people involved, I am unable to list each name individually, but I would like to make particular mention of Messrs. Chen De Quan and Ji Lei, who have been with me since the inception of my work up to the present time.

A part of the book has been written by Professor Wang Yuan, based on my report. In view of my own heavy work load and poor health, this book could not have been completed so quickly had he not participated in the writing of this book.

Finally, I must also thank Drs. K. Peters and W. Klump of Birkhäuser Boston, Inc., for their help during the production of this book.

December, 1982

Hua Loo-Keng

1. This report was also presented in England at Birmingham University and the London Mathematical Society, and at the American Mathematical Society and various universities in Western Europe and America.

Preface

Professor Hua Loo-Keng is the first person to have undertaken the task of popularizing mathematical methods in China. As early as 1958, he proposed that the application of operations research methods be initiated in industrial production. With his students, Yu Ming-I, Wan Zhe Xian and Wang Yuan, Professor Hua visited various transportation departments to promote mathematical methods for dealing with transportation problems, and a mass campaign was organized by them and other mathematicians to advance and apply linear programming methods to industrial production in Beijing and in Shandong province. However, due to the fact that these methods have limited applications and their computation is rather complex, their popularization and utilization in China have so far been restricted to a small number of sectors such as the above mentioned transportation departments.

In 1958 Hua Loo-Keng proposed the use of Input-Output methods in the formulation of national economic plans. Apart from publicizing this method, he carried out in-depth research on the subject. He also gave lectures on related non-negative matrix theory, pointing out the economic significance of various theoretical results.

The actual popularization of mathematical methods in China's industrial sector was restarted in 1965. Hua Loo-Keng made an incisive summary of his experiences gained in this field in the preceding years, stating the objective and substance of promulgating mathematical methods and identifying ways in which such methods should be implemented. He refers to this as the "Three Principles", namely 1) For Whom? 2) Which Technique? and 3) How to Popularize? (See the Introduction). A team was formed with Hua Loo-Keng as head and his students, Chen De Quan and Ji Lei, as the core members. As topics for dissemination, Hua chose the Optimum Seeking Methods that may be used in problems of improving production technology and the Overall Planning Methods, that may be applied in problems of production organization and management. The task of sifting through an extensive body of literature to identify mathematical methods suitable for application and popularization in a country with numerous small factories and a relatively backward industrial technology, as well as explaining these methods in a language comprehensible to the vast and relatively poorly educated workforce was, in itself, extremely difficult. Consequently, Hua Loo‑Keng wrote two general scientific readers, "Popular Lectures on Optimum Seeking Methods" (with Supplements), and, "Popular Lectures on Overall Planning Methods" (with Supplements), both of whom were very well received. Altogether, the amount of time and energy expended by Hua Loo-Keng in popularizing mathematical methods has been no less than his efforts in pioneering research in certain areas of pure mathematics. Over the last twenty years, Hua has managed to reserve time to continue his work on pure mathematics, and has always earnestly researched into the theory and striven for a rigorous proof of each method that he has handled in the popularization work.

Since 1965, Hua Loo‑Keng and his team have travelled to several hundred cities in more than twenty provinces, municipalities and autonomous regions, visiting thousands of factories and giving lectures to several million workers and professionals, helping them to understand the Optimum Seeking Methods and the Overall Planning Methods and also to apply these methods in order to improve their work. For instance, upon the team's arrival at a factory or a mine, some two hundred people from within and outside the province are brought together to attend a one week training course. Apart from a number of lectures, most of the time is spent in small group discussions, the subject of discussion being mainly how to apply the Optimum Seeking Methods and the Overall

Planning Methods to either one's own work or to the local production process. At the end of the week, the team travels to various industrial units within the province to work together with the factory leader, workers and professionals. During this time, Hua Loo-Keng himself invariably goes to the major cities and factories in the province to provide personal guidance and familiarize himself with their experiences, whether successful or unsuccessful. A meeting with representatives from other cities is then convened at a specific factory or mine to learn from these experiences, so that they can improve their own work. Thus, the scope and domain of the team's operations are extensive. Solutions to as many as 10,000 problems can often be provided in one province alone, creating considerable economic benefits. At the same time, prior to departure, the team ensures that the local people are able to independently apply the Optimum Seeking Methods and the Overall Planning Methods to their work. It goes without saying that the excellent results achieved in the propagation of mathematical methods in China contributed to the tremendous prestige that Hua Loo Keng enjoys amongst the Chinese people.

In 1980, at the invitation of Dr. H. O. Pollak, Hua Loo-Keng addressed the Fourth International Conference on Mathematical Education and the Bell Laboratories on "Some Personal Experiences in Popularizing Mathematical Methods in the People's Republic of China". A revised version of this paper forms the introduction to the present book. The various topics raised in the introduction are elaborated on in the subsequent chapters, the writing of which I myself was involved in. The material on Input-Output analysis was too extensive and has therefore not been included in this book.

I have been doing research on number theory with Professor Hua since 1952. However, I have only participated in the work on the popularization of mathematical methods on a few occasions and as a result certain omissions and errors in my writing are unavoidable. Since the scope of this book is broad and the literature diverse, it is only possible to cite a limited number of references.

December, 1982

Wang Yuan

Preface by the editor

Four types of changes have been made to the original manuscript.

1. Due to language difficulties, the entire manuscript has been rewritten. The revisions were a compromise between the need for English language clarity and the goal to stay close to the atmosphere of the manuscript. There are some passages where the text is still ambiguous; the editor was unable to discern the intended meaning and has therefore opted to keep the ambiguities.

2. Some changes have been made in formulations of theorems, in proofs of theorems and in examples, in order to enhance the clarity. All such changes are within the confines of the originally presented ideas.

3. The decision was made to replace the original Chapter 1, "Approximate Calculation of Surface Area and Volume of a Body" by the more extensive paper "On the Calculation of Mineral Reserves and Hillside Areas on Contour Maps", by Hua Loo-Keng and Wang Yuan, published in Acta Math, Sinica, 11 (1961), and translated in English by Hu Dide. The paper was published in "Loo-Keng Hua, Selected Papers", edited by H. Halberstam, Springer Verlag 1983.

4. The decision was made to drop the very short Chapter 12, "Generalised Inverse of Matrices", as well as some references to it in Chapter 0, since the topic did not seem to fit in with the rest of the book. For the same reason, a section of Chapter 2, "Launch Point and Range of a Rocket", was omitted.

Since the original manuscript did not contain many recent and easily accessible references, the editor has added some standard references at the end of various chapters.

Hua Loo-Keng died in Tokyo on June 12, 1985. Included here is "An Obituary of Loo-Keng Hua" by Heini Halberstam, previously published in The Mathematical Intelligencer, Vol. 8, No. 4, and generously made available by Springer Verlag.

The following are some references on the state of affairs of applied mathematics in the People's Republic of China:

Fitzgerald, Anne, and Saunders MacLane, (eds.), Pure and Applied Mathematics in the People's Republic of China: A Trip Report of the American Pure and Applied Mathematics Delegation. National Academy of Sciences, Washington D.C. 1977.

Gina Bari Kolata, "Hua Lo-Keng Shapes Chinese Math", Science, Vol. 210, 24 October 1980.

"MS/OR in China", Interfaces, 16:2, March-April 1986. (This includes the article "Popularization of Management Science in China", by Chen Dequan, Ji Lei and Pang Chungmin on the work of Hua Loo-Keng and his group.)

Finally, I wish to thank Dr. W. F. Lucas and Dr. H.O. Pollak for valuable discussions and suggestions on the editing of this manuscript, and Ms. Lauren Cowles from Birkhäuser Boston Inc. for her help and her infinite patience with this project.

August 1988

J.G.C. Heijmans

In a letter dated June 19, 1984, Loo-keng Hua wrote: "I wish there were enough time left to me to return to the university to give a course, to make me think again about the '40s which were the golden years of my life." It was not to be; he died of a heart-attack while giving a lecture in Tokyo on June 12, 1985, full of honors, a hero to his people, doing what he had done so well all his life—expounding his ideas.

Hua's life has all the ingredients of a modern legend. I do not know the exact date of his birth, and even the year of his birth has some uncertainty attached to it. He was probably born in November, 1909, into a poor family residing in Jinton County, in the Jiangsu Province of China. He had no more than nine years of formal schooling, yet he wrote his first paper, on Sturm's theorem, at the age of nineteen. Soon afterwards he became assistant and later lecturer at Tsinghua University in Beijing, and then the China Foundation awarded him a research fellowship for several years. Norbert Wiener had visited China in the mid '30s and must have been impressed by Hua; he reported the encounter to G. H. Hardy and in 1936 Hua arrived in Cambridge for a two-year visit. This was a decisive experience for the young Hua. While his earliest publications had shown his mathematical interests to be wide-ranging, his efforts had begun to focus on Waring's Problem as early as 1934 and during the Cambridge period he laid the foundations for his enduring contributions to additive number theory. It is quite astonishing to realize that these two years, then three important months during 1945–46 with I.M. Vinogradov in Russia and barely five years in the USA, from 1946–1950, were all the time that Hua spent at major mathematical centers in the west (three years at Princeton and two at Illinois); yet during this period he embarked on his researches in matrix geometry, in functions of several complex variables, equations over finite fields, automorphisms of symplectic groups, while yet he continued to make fundamental contributions to analytic number theory.

Hua stood in the grand tradition of scholar-teachers. Possessed of a commanding presence, a magnetic personality and a romantic imagination, he was not only a superb director of research but an outstanding teacher at every conceivable level of instruction. Several of his scholarly treatises are still in current use the world over even though they were written many years ago, after his final return to China in 1950, to guide his research students; but these books are only a small fraction of all that he wrote, with prodigious industry and quenchless enthusiasm, to teach his people about mathematics. Much later, when he was called upon to act as a kind of mathematical trouble shooter in his country's drive towards industrialization, he travelled with a support team of scientists the length and breadth of China to show workers how to apply their reasoning faculty to a multitude of everyday problems. Whether in ad hoc problem-solving sessions on the shopfloor or in huge open-air meetings outside factories, he touched the multitudes with the spirit of mathematics as no mathematician had ever done before, anywhere in the world. Groping for western analogies, one might think of Abelard filling the halls of the Sorbonne in Paris, or of John Wesley's preaching in England and America; but perhaps this vast surrender to the influence of a popular teacher is really an eastern phenomenon which we in the west cannot ever quite fathom. We are given just a glimpse of the scale of Hua's consulting activities by the list of 152 themes at the end of the "Selecta": [1]—even the general areas of application are too numerous to list here. Hua talked of one day publishing the details of these investigations, but I fear now that this will never happen; we may have to be satisfied with the article "on the calculation of mineral reserves and hillside areas on contour maps" (with Wang Yuan) which was translated from the Chinese specially for the "Selecta".

Even at the level of convential scholarship one can follow Hua's move towards applications of mathematics from the late '50s onward, in his later publications on topics in differential equations, operations research and numerical integration. In the last of these, however, Hua's scientific opus comes elegantly to the

completion of a full circle: ideas from uniform distribution and algebraic number theory are used, in joint
work with Wang, as an alternative to the Monte Carlo
method in numerical quadrature of multiple integrals.
The last time he lectured in my department, in 1984,
he gave a lively talk, in a packed lecture room, on
mathematical economics.

The years of the "Cultural Revolution" were hard
for Hua and he attributed his survival, at least in part,
to the personal protection of Mao. Even so, many of
his papers and manuscripts were confiscated and presumably destroyed. After the thaw, Hua was sur-
prised to receive permission to accept an invitation
from Prof. Livingstone to visit the University of Birmingham, England, on a senior visiting fellowship
from the then Science Research Council of the UK; and
Hua visited Europe and the USA on several subsequent occasions. In many respects these visits were
the personal triumph his romantic nature had longed
for after 30 years of isolation, and one could gauge
from the way in which Chinese communities every-
where and of all political persuasions flocked to meet
him, or even just to see him, the status he had attained in China. (It seems that Chinese television
made a six-hour mini-series of the early years of his
life.) He was eager to renew mathematical acquain-
tances in his youth; at the same time it was poignant
to witness his sadness at the passing of many others
he would dearly have loved to meet once more.

If Hua regretted leaving the US when at the height
of his powers and felt keenly, when later he revisited
the West, that he could not recover what the years had
lost, his commitment to his own country was absolute
and unswerving. He returned to China in '38 and
served until 1945 as professor at the Southwest Associated University in Kunming (in the Province of
Yunan), an institution formed by the merging of several universities dispossessed by the Japanese inva-
sion of China. And again, after the creation of the
People's Republic of China in '49 he returned the following year to the mainland to help in the rebuilding
of the Mathematical Institute of the Academia Sinica.
He became Director of the Institute in 1952 and re-
mained in this post until 1984, when he relinquished it

to Prof. Wang. He was for many years also Vice-President of Academia Sinica and Rector of the University for Science and Technology, Hefei (Province Anhwei). In 1957 he received a Prize of the Academia Sinica, for his work on "Harmonic analysis of function of several complex variables in classical domains". In 1980 he received an Honorary Doctorate from the University of Nancy, France, in 1983 he received another from the Chinese University of Hong Kong, and in 1984 a third from the University of Illinois. In 1982 he became a foreign associate of the U.S. Academy of Sciences, in 1983 he became a Member of the Academy of the Third World, and in 1985 he was elected a member of the German Academy of Bavaria.

CHAPTER 0

INTRODUCTION

§ 0.1 Three principles.

Since the middle of the sixties, my assistants and I have visited twenty-six provinces, municipalities and autonomous regions in China, hundreds of cities and thousands of factories, and we have met with millions of workers, peasants and technicians. The following questions sum up our experiences accumulated in this still on-going endeavor:

1) "To whom?" and "For what purpose?"

2) "Which technique?" and

3) "How to popularize?"

Some elaboration of these three questions is now in order.

1) "To Whom?" Specialists and workers do not often share a common language. My experiences have shown me that in order to achieve a common language we must have common interests. It is hard to imagine that a worker who is busily occupied balancing a grinding wheel or a cylinder would be interested in talk about infinite dimensional spaces even though infinite dimensional spaces are interesting and beautiful to mathematicians. Therefore, to begin with popularization, there must be some common goal that is interesting to both the lecturer and the audience. With such a common goal, they can speak in a common language. Only after accomplishing this we can then talk about question 2).

2) "Which technique?" I will give a somewhat detailed picture later about the techniques we have chosen. For the moment I only want

1

to mention the "Three Principles" for selecting techniques for popularization.

a) Popularity. All methods we provide should be easy to understand, easy to use, and easily obtain effective results.

b) Practicality. A method, before being popularized, should be tested in order to determine to what areas it can be applied. Then it could be popularized within those areas. In practice, we have found that it is not always successful to apply a method in our country even though it has been successful in other countries.

c) Theoretically sound. This is necessary for introducing profound knowledge in an easy way, for discriminating the best method from many others, and for creating new methods.

3) "How to popularize?" It is my experience that we should go to factories in person and start work on a small scale project, for example in a workshop. If our suggestions turned out to be effective, they would then naturally attract wider attention and we might soon be invited to other workshops. If most of the supervisors and workers in this factory became interested in the methods we presented, these methods could be extended to the whole factory or even to a whole city or a whole province. In this way, my assistants and I sometimes gave lectures before an audience of more than a hundred thousand! Of course, the audience had to be divided into smaller groups. Since closed television circuits had not yet become available in our country, my assistants were sent to these various groups to give the necessary demonstrations and graphs while the audience was listening to my lecture being given in the main hall through a sound system. After the lecture, we not only answered questions asked by the audience, but also we went to the workshops to work together with workers applying the methods to their projects and to improve their techniques.

§ 0.2 Looking for problems in the literature.

A researcher often finds his methods for solving his problem in books or the literature. I think it would be a good approach to do research work if one would analyze and compare the different methods that are available. There are many successful examples of this approach and I will cite only one of them as an illustration.

How to calculate the surface area of a mountain? We found two methods for solving this problem in books. One is the so-called Bayman's method which is often used by geologists. The other is the so-called Volkov's method often used by geographers. These two methods can be described briefly as follows.

Let us start with a contour map of a mountain in which the difference of heights between two neighboring contours is Δh. Let l_0 denote the contour of height 0, l_1 the contour of height Δh, l_r the contour of height $r \times \Delta h$, and let l_n correspond to the highest point of the mountain of height h. Also, let B_i denote the area on the map between l_i and l_{i+1}.

1) Bayman's method:

a) Compute

$$C_i = \tfrac{1}{2}(|l_i| + |l_{i+1}|)\Delta h,$$

where $|l_i|$ denotes the length of the contour l_i.

b) Let

$$Ba_n = \sum_{i=0}^{n-1} \sqrt{B_i^2 + C_i^2}.$$

The geologists regard Ba_n as the approximate surface area of the mountain.

2) **Volkov's method**

a) Let

$$I = \sum_{i=0}^{n-1} |I_i| \qquad and \qquad B = \sum_{i=0}^{n-1} B_i .$$

b) Let

$$Vo_n = \sqrt{B^2 + (I\Delta h)^2} .$$

The geographers regard Vo_n as the approximate surface area of the mountain.

These are two methods we found in different branches of science. We may immediately raise two obvious problems, namely, 1) do these methods converge to the real surface area? and 2) which method is better?

Unfortunately, both of these methods do not converge to the real surface area S. To be exact, letting

$$Ba = \lim_{n\to\infty} Ba_n , \qquad Vo = \lim_{n\to\infty} Vo_n ,$$

it turns out that

$$Vo \leq Ba \leq S.$$

The proof seems to be interesting although quite simple. We use cylindrical polar coordinates (ρ, θ, z), where the highest point is chosen to be the origin. We denote the surface by the equation

$$\rho = \rho(z, \theta), \qquad 0 \leq \theta \leq 2\pi,$$

which is also the equation of the contour of height z. It is well-known that

$$S = \int_0^h \int_0^{2\pi} \sqrt{\rho^2 + (\frac{\partial \rho}{\partial \theta})^2 + (\rho\frac{\partial \rho}{\partial z})^2} \; d\theta dz.$$

Introducing the complex valued function

$$f(z, \theta) = -\rho\frac{\partial \rho}{\partial z} + i\sqrt{\rho^2 + (\frac{\partial \rho}{\partial \theta})^2},$$

we may write

$$\text{Vo} = \left| \int_0^h \int_0^{2\pi} f(z, \theta) \, d\theta \, dz \right| \le \text{Ba} = \int_0^h \left| \int_0^{2\pi} f(z, \theta) \, d\theta \right| dz$$

$$\le S = \int_0^h \int_0^{2\pi} | f(z, \theta) | \, d\theta \, dz.$$

Unfortunately, the equalities hold only in some very special cases. This result shows that neither method can give a good approximation of the surface area. However, it points a way towards a more satisfactory approximation. This example clearly demonstrates the importance of a sound theoretical analysis. However, in view of the fact that the above problem is of interest to only a very limited audience, it is certainly not a good subject for popularization. We can accumulate such materials in our textbook to make its contents richer and more practical.

This example shows us the possibility of finding mathematical research problems in other fields of science. (See Chapter 1).

§ 0.3 Looking for problems in the workshop.

Many problems found in workshops or raised by workers are very interesting. Here I give a meshing gear-pair problem as an example.

In 1973 we visited the city Loyang in central China to popularize optimum seeking methods and overall planning methods. A worker in the Loyang tractor factory asked us a question about meshing gear-pairs that could be stated mathematically as follows.

Given a real number ξ, determine four integers a, b, c and d between 20 and 100 such that

$$\left| \xi - \frac{a\,b}{c\,d} \right|$$

is a minimum.

This worker pointed out to us that the numbers recommended in a handbook for engineers were not always the best ones. He took $\xi = \pi$ as an example. The number given by the handbook is

$$\frac{377}{120} = \frac{52 \times 29}{20 \times 24}.$$

He himself had found the number

$$\frac{2108}{671} = \frac{68 \times 62}{22 \times 61},$$

which is a better approximation than the number in the handbook. He then raised the question, "Are there better approximations?"

This is a problem of diophantine approximations. It seemed to be easy to solve by means of the theory of continued fraction. One may expect that one of the convergents of π , i.e.

$$\frac{3}{1}, \ \frac{22}{7}, \ \frac{333}{106}, \ \frac{355}{113}, \ \frac{103993}{33102}, \ \cdots$$

would be better than his number. But this is not the case! The number

$\frac{2108}{671}$ is better than $\frac{3}{1}$, $\frac{22}{7}$ and $\frac{333}{106}$, and all the other convergents of π can not be expressed in the required from. Therefore, the convergents of π are all inappropriate. Thus it is clear that a direct application of the continued fraction approximation is futile. What can we do?

There was only one day left! On the day of my departure from Loyang, I gave my assistants a small piece of paper on which I wrote

$$\frac{377}{120} = \frac{22 + 355}{7 + 113}$$

They understood my suggestion to attack the problem with the method of Farey means. In this way, they found two approximations that are better than the one proposed by the worker, namely,

$$\frac{19 \times 355 + 3 \times 333}{19 \times 113 + 3 \times 106} = \frac{7744}{2465} = \frac{88 \times 88}{85 \times 29}$$

and

$$\frac{11 \times 355 + 22}{11 \times 113 + 7} = \frac{3927}{1250} = \frac{51 \times 77}{50 \times 25}.$$

The latter is the best approximation.

Of course, this method is applicable for any real number ξ. My assistants and the factory workers re-examined the handbook carefully and replaced a number of recommended ratios by better ones (practically the best possible ones). They even filled in a few cases left blank in the textbook.

But this technique was still not sufficiently general for popularization. Only very few workers may encounter this problem in their work. Also, if the handbook is improved using this method, it will serve the same purpose (See Chapter 2).

The problem of how to choose proper techniques is still left to be studied.

§ 0.4 Optimum seeking methods (O.S.M.).

The "method of trials by shifting to and fro" is an often used method. But how can we do it effectively? It was J. Kiefer (1953) who solved this problem first. Kiefer's method is called the "golden section method" because it is connected with the golden number. This is a widely used method. How can we make ordinary workers understand it well and then use it in their work?

I now describe some experiences in this respect. I usually introduce this method with the following steps using a strip of paper.

1) First the audience is asked to memorize the number 0.618.

2) As an example we consider a technological process for which the best temperature lies somewhere between 1000°C and 2000°C. We may perform one thousand experiments at 1001°C, 1002°C, ... , and then we are bound to find the best temperature. But a total of one thousand experiments are too much for a factory.

3) (Showing the strip of paper). Suppose that this is a strip of paper with a scale from 1000°C to 2000°C on it. The first trial point is taken at 0.618 of the whole range, i.e., at 1618°C (See Figure 0.1). I often use a cigarette to scorch a hole on the strip. The result of this first trial is then recorded.

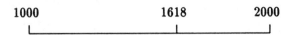

1000   1618   2000

Figure 0.1

4) Then I simply fold the strip in half. The second trial point is taken at the point opposite the first point, i.e. at 1382°C (See Figure 0.2).

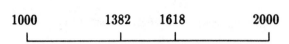

Figure 0.2

Then the results at these two points are compared and I tear the strip at the worst point and keep the part containing the better point.

5) Step 4) is repeated for the remaining strip until the best point is reached.

In this way many ordinary Chinese workers, usually with rather inadequate formal education, can understand the golden section method quickly and get a good mastery of it. In addition to this example I have chosen several often used methods according to the "Three Principles" of selecting topics mentioned in section 0.1. By adopting similar visual and intuitive approaches, we have popularized them with gratifying results.

We have seriously studied some methods which may be used in special cases, even though they are difficult for popularization. For example, the rate of convergence for the DFP technique (1962) was originally claimed to be

$$\| \vec{x}^{k+1} - \vec{x}^* \| = o\left(\| \vec{x}^k - \vec{x}^* \|\right).$$

We proved in the mid sixties that the rate of convergence for this technique can be

$$\| \vec{x}^{k+s} - \vec{x}^* \| = O\left(\| \vec{x}^k - \vec{x}^* \|^2\right).$$

When I visited Western Europe in 1979, I was told that this result was also proved by W. Burmeister in 1973.

In 1968 we developed a technique with rate of convergence

$$\| \vec{x}^{k+1} - \vec{x}^* \| = \sigma\left(\| \vec{x}^k - \vec{x}^* \|^2 \right).$$

Using this technique, the number of necessary experiments is at most one half that of the DFP technique (See Chapters 3 and 4).

§ 0.5 The Fibonacci search.

Sometimes the parameters in a technological process are not allowed to vary continuously. A lathe, say, has only a limited number of turning speeds so that it is difficult to apply the number $\frac{\sqrt{5}-1}{2} \approx 0.618$. In this case the theory of continued fractions can play an important role. The golden number $\frac{\sqrt{5}-1}{2}$ has convergents

$$\frac{0}{1}, \; \frac{1}{1}, \; \frac{1}{2}, \; \frac{2}{3}, \; \frac{3}{5}, \; \frac{5}{8}, \; \frac{8}{13}, \; \cdots , \; \frac{F_n}{F_{n+1}}, \cdots ,$$

where the F_n's are the Fibonacci numbers which are defined by $F_0 = 0$, $F_1 = 1$ and $F_{n+2} = F_{n+1} + F_n$ $(n \geq 0)$. Besides the example of a lathe, sometimes I illustrate the Fibonacci search with matches.

Suppose a lathe has twelve speeds

$$\boxed{1} \; \boxed{2} \; \boxed{3} \; \boxed{4} \; \boxed{5} \; \boxed{6} \; \boxed{7} \; \boxed{8} \; \boxed{9} \; \boxed{10} \; \boxed{11} \; \boxed{12}$$

and we use twelve matches to represent them. We suggest that the first trial should be made at the eighth speed and the second trial at the fifth speed which is symmetric to the eighth. Then the results of these two trials are compared. If $\boxed{8}$ gives the better result, then we remove five matches starting from the left and keep

$$\boxed{6} \; \boxed{7} \; \boxed{8} \; \boxed{9} \; \boxed{10} \; \boxed{11} \; \boxed{12}.$$

Otherwise, remove the five rightmost matches and we have

$$\boxed{1}\ \boxed{2}\ \boxed{3}\ \boxed{4}\ \boxed{\dot{5}}\ \boxed{6}\ \boxed{7}.$$

In either case, seven matches remain.

Again using symmetry, we make the next trial at $\boxed{10}$ if $\boxed{8}$ gave the better result. If $\boxed{8}$ is still the better one, we remove $\boxed{10}$ $\boxed{11}$ and $\boxed{12}$ and keep $\boxed{6}\ \boxed{7}\ \boxed{8}\ \boxed{9}$.

Then, by symmetry, we make the next trial at $\boxed{7}$. Assuming the result at $\boxed{7}$ is better than that at $\boxed{8}$, we are left with

$$\boxed{6}\ \boxed{\dot{7}}.$$

Finally we make the last trial at $\boxed{6}$. If $\boxed{6}$ is better than $\boxed{7}$, then $\boxed{6}$ is the best speed among the twelve speeds.

Such a method can be mastered without difficulty by turners (See Chapters 3 and 4).

§ 0.6 The golden number and numerical integration.

The number $\dfrac{\sqrt{5}-1}{2}$ is called the golden number. It is not only useful for the golden section, but also plays an important role in the theory of diophantine approximation. It inspired us to think of the following formula for numerical integration:

$$\int_0^1\int_0^1 f(x,y)\,dx\,dy \approx \frac{1}{F_{n+1}}\sum_{t=1}^{F_{n+1}} f\left(\frac{t}{F_{n+1}}, \left\{\frac{F_n t}{F_{n+1}}\right\}\right),$$

where $\{\xi\}$ is the fractional part of ξ. This is a formula which approximates a double integral by a single summation.

How can we extend this formula to the multi-dimensional case? The idea is that we should recognize what $\frac{\sqrt{5}-1}{2}$ really is. It is obtained by dividing the unit circle into five equal parts, i.e. by solving the equation

$$x^5 = 1$$

or

$$x^4 + x^3 + x^2 + x + 1 = 0.$$

Letting

$$y = x + x^{-1},$$

we have

$$y^2 + y - 1 = 0,$$

giving

$$y = \frac{\sqrt{5}-1}{2}$$

or

$$y = 2\cos\frac{2\pi}{5}.$$

This is a cyclotomic number. Since the number $\frac{\sqrt{5}-1}{2}$, obtained by dividing a circle into five parts, is so useful, it is natural to think that the numbers

$$2\cos\frac{2\pi l}{p}, \qquad 1 \le l \le \frac{p-1}{2} = s$$

which are obtained by dividing the unit circle into p equal parts, might be useful for numerical integration of multi-dimensional integrals, where p denotes a prime number ≥ 5.

By Minkowski's theorem, we may confirm that there exist x_1, x_2, ..., x_{s-1} and y so that

$$\left| 2\cos\frac{2\pi l}{p} - \frac{x_l}{y} \right| \le \frac{s-1}{s\,y^{\frac{s}{s-1}}}, \qquad 1 \le l \le s-1.$$

However, Minkowski's proof only concerns the existence of $x_1, x_2, \dots,$ x_{s-1} and y. As for the cyclotomic field $Q(\cos\frac{2\pi}{p})$, we can find $x_1, x_2,$ $\dots, x_{s-1}$ and y effectively since we have an explicit independent unit system. Thus, we may replace

$$\left(\frac{t}{F_{n+1}}, \left\{\frac{F_n\,t}{F_{n+1}}\right\}\right), \qquad t = 1, \dots, F_{n+1}$$

by

$$\left(\frac{t}{y}, \left\{\frac{x_1\,t}{y}\right\}\right), \dots, \left\{\frac{x_{s-1}\,t}{y}\right\}\right), \qquad t = 1, \dots, y.$$

This sequence of points may be used not only in numerical integration but also when uniformly distributed quasi-random numbers are being used (See Chapter 5).

§ 0.7 Overall planning methods (O.P.M.)

One of the purposes of education reform is to increase the practical knowledge of students so that when they later go on to the practical world they may apply their knowledge to improve production processes, engineering designs or any other kind of work they may be involved in. The Optimum Seeking Methods presented above are good methods for this purpose since they can be used to improve production technologies, and at the same time they are easy to popularize.

We know that quality control methods can prevent unqualified products from being sent out of the factory so that the reputation of the factory can be retained. But rather than disposing products of poor quality after they are made, it would be better if we used Optimum

Seeking Methods beforehand to find the best operating conditions in order to reduce the rate of the unqualified products.

In addition to problems of production technology, there are also problems of management and of organization of production. To deal with these problems, some effective mathematical methods may be used. We call them Overall Planning Methods (O.P.M.).

O.P.M. consist of many methods. Some of them are very useful and can be easily popularized. I will mention a few examples.

1) Critical Path Method (CPM).

Since CPM is well known, I will only describe how we (work on) popularize it rather than give the details of the method. When we start with popularizing CPM, we often set as our objective the minimization of the total time duration for a certain project. If the essential idea of the method has been mastered, more sophisticated problems such as minimizing the production cost or allocating manpower and other resources can be handled without any difficulty.

Our primary goal is to explain the CPM by solving some practical projects so that the technicians and workers can learn it within their own work situation. The steps are:

a) Investigation. Three items are to be investigated: i) a list of all activities relevant to the project; ii) the relations between these activities, and iii) the duration of each activity. In order to do this well, we must rely on the technicians and workers involved in the related activities because their estimates are certainly more reliable than those made by others.

b) Critical Path diagram. Based on the above information, the technicians and workers learn how to draw drafts and how to find all critical paths. Then a general discussion is held to see if there is any possibility that the total time duration for that project can be further shortened. In this way a Critical Path diagram is determined and a plan for the project is scheduled.

c) Execution of the CP diagram. During the execution the CP diagram has to be constantly reviewed and adjusted since some activities may have taken more time or less time than estimated. Thus, the critical paths may be changed at times.

d) Summing up. At the completion of a project, the actual stage-by-stage progress is recorded in the form of a CP diagram so that this information may be utilized for future projects of a similar type.

We have found that CPM can be effectively applied to small scale projects as well as very large scale projects. We may start with drawing a CP diagram for every sub-project and then put all sub-diagrams together to get an overall CP diagram for the whole project. Or reversely, we may first make a rough plan for the whole project which directs all subunits to draw their own CP diagrams and then these diagrams are synthesized into a general diagram that may then be discussed and modified (see Chapter 6).

2) Sequencing analysis.

Suppose there are a number of projects to be executed and these projects should be carried out in sequence without restrictions on the order. How can the tasks be arranged so that the total waiting time is a minimum?

A mathematical problem relevant to this question is the following. Given two sets of non-negative numbers

$$a_1, a_2, \ldots, a_n,$$

and

$$b_1, b_2, \ldots, b_n,$$

which permutations (a_{s_i}) and (b_{t_i}) will minimize (maximize) the sum

$$\sum_{i=1}^{n} a_{s_i} b_{t_i} ?$$

The answer is that the maximum is obtained by ordering (a_i) and (b_i) according to size, and the minimum is obtained by ordering the sequences in opposite order. The proof is simple. In order to prove the general statement we need only observe the following simplest case:

If

$$a_1 \leq a_2, \qquad b_1 \leq b_2$$

then

$$a_1 b_1 + a_2 b_2 \geq a_1 b_2 + a_2 b_1,$$

that is

$$(a_2 - a_1)(b_1 - b_2) \geq 0.$$

In general, if two a terms have the same order as the two corresponding b terms, then we can decrease the sum by reversing the order in one of the two sequences.

After explaining the above mathematical problem, I would often talk about sequencing analysis using a simple example. A water tap is used to fill n buckets. If the capacities are not the same, how can the buckets be arranged so that the total waiting time is a minimum?

Suppose a_1 is the time needed to fill the first bucket, a_2 for the second bucket, ... , and a_n for the n-th bucket. Then, the first bucket has to wait the amount of time a_1, the second bucket has to wait the amount of time $a_1 + a_2$, the third for $a_1 + a_2 + a_3$ and so on. The total waiting time is given by

$$T = a_1 + (a_1 + a_2) + \ldots + (a_1 + a_2 + \ldots + a_n)$$

$$= n a_1 + (n-1) a_2 + \ldots + a_1,$$

which is minimized when the sequence a is arranged in ascending order, i.e.

$$a_1 \le a_2 \le \cdots \le a_n .$$

That is to say, the smallest buckets should be filled first.

Suppose that there are s water taps, and there are m buckets of capacities $a_1^1, a_2^1, \ldots, a_m^1$ for the first tap, m buckets of capacities a_1^2, $a_2^2, \ldots, a_m^2$ for the second tap, etc. (If the taps are assigned different numbers of buckets, the numbers can still be assumed to be equal by introducing appropriate numbers of buckets of zero capacity.) The total waiting time T is given by

$$T = \sum_{j=1}^{s} (m\,a_1^j + (m-1)a_2^j + \cdots + a_1^j).$$

On setting

$$b_1 = b_2 = \cdots = b_s = m$$

$$b_{s+1} = b_{s+2} = \cdots = b_{2s} = m - 1,$$

etc., the same result is obtained, namely start with the smaller buckets in order to minimize T. We do not have many problems connected with sequencing analysis. In general, these problems are too complicated to be solved by standard mathematical methods. Of course, in such cases only partially satisfactory solutions are obtained (see Chapter 8).

3). The graphical method on the transportation problem is another method which can be popularized. Suppose there are m shipping origins $A_1, \ldots, A_m$ which produce $a_1, \ldots, a_m$ tons of wheat respectively, and that there are n shipping destinations (consumer places) $B_1, \ldots, B_n$ that require $b_1, \ldots, b_n$ tons of wheat respectively. The objective is to minimize the total amount of ton-kilometres of transported goods. This problem can of course be solved by linear programming methods. But we often use a simpler, more intuitive method, the graphical method. The idea behind this method is to draw a flow diagram of the shipped goods on a map so that there are no counter flows, and so that on each loop the total length of the flow in one direction is no more than half the length of the whole loop. It can be proved that such a flow diagram represents an optimal solution to the transportation problem (See Chapters 9, 10 and 11).

§ 0.8 On the use of statistics.

1). Empirical formulas and the importance of mathematical insight.

Sometimes an empirical formula is derived from a set of statistical data. The meaning of such a formula can often be recognized by scientists with a good sense and some mathematical training as in the following example. R.C. Bose, a mathematical statistician from India, obtained the empirical formula

$$A = \frac{\text{length} \times \text{width}}{1.2}$$

for the area A of a leaf of rice grain, using a large sample of such leaves in India. I had no reason to question its reliability. However, some agriculturists in China applied this formula to the harvest on their rice farm. On seeing the shape of the leaves on their farm, I pointed out to them that this formula could not be suitable for their leaves. They took some samples of their leaves and discovered that this formula tends to overestimate the areas. After their expression of surprise, I gave them a simple explanation by drawing the following picture, where the shaded region represents the leaf (See Figure 0.3).

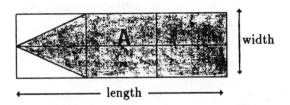

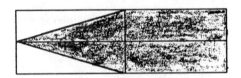

Figure 0.3

In this case, the ratio of the area of the rectangle to A is $\frac{6}{5}$, i.e. 1.2. In contrast to this, a typical leaf on their farm has a more elongated tip and I drew another picture (See Figure 0.4). Thus, in this case, the ratio is $\frac{4}{3}$ which is naturally larger that 1.2. This explains the overestimation of A by using Bose's formula for their farm.

Figure 0.4

We all learned the lesson on the importance of seeking mathematical insight beneath an empirical formula.

2) A simple statistical method.

In the experimental sciences we frequently use statistical methods. Needless to say, their importance must not be denied. However, it is my view that some methods tend to be too sophisticated and can easily be misused. I give a few examples first.

Example 1. Let $x_1, \ldots, x_{20}$ denote twenty observations from independent repetitions of an experiment. Let

$$\bar{x} = \frac{x_1 + \cdots + x_{20}}{20} \quad \text{(the sample mean)}$$

and

$$s = \sqrt{\frac{\sum_{i=1}^{20} (x_i - \bar{x})^2}{19}} \quad \text{(the sample standard deviation)}.$$

The experimenter might then claim that the interval

$$(\bar{x} - \frac{1.96\,s}{\sqrt{20}}, \bar{x} + \frac{1.96\,s}{\sqrt{20}})$$

is a "reasonable" estimate of some underlying "average". Such a sophisticated method does not seem to be easily understood by an ordinary worker in China. Besides, the underlying Gaussian assumption might not be valid in his case!

In practical terms, I tend to prefer the following approach. Let

$$x_{(1)} \leq x_{(2)} \leq \cdots \leq x_{(20)}$$

denote an ordered sample of observations. We might safely say that the possibility of an observation falling in the interval

$$[\frac{x_{(1)} + x_{(2)}}{2}, \frac{x_{(19)} + x_{(20)}}{2}]$$

is more than $\frac{18}{20} = 0.9$. This method may be more readily accepted by our factory workers and farmers.

Example 2. Suppose that it is required to decide which of two production methods is better, based on five observations from each method. Let $(a_1, \ldots, a_5)$ and $(b_1, \ldots, b_5)$ denote the samples of the first and second method respectively. It may be tempting to simply compare the means of the two samples by referring to the usual Student's t distribution. Underlying such a sophisticated method are a number of assumptions, such as normality, equal variance and independence, which are not easily understood by ordinary factory workers in China.

It seems to me that a more robust and simple method based on the ordered samples $a_{(1)} > a_{(2)} > \ldots > a_{(5)}$ and $b_{(1)} > b_{(2)} > \ldots > b_{(5)}$ may be more suitable for popularization in China. For example, an ordinary factory worker can accept it as being sensible to reject the hypothesis H_0 that the two methods are equally good in favor of the first method if the mixed order sample is either

$$a_{(1)} > a_{(2)} > a_{(3)} > a_{(4)} > b_{(1)} > a_{(5)} > b_{(2)} > b_{(3)} > b_{(4)} > b_{(5)}$$

or

$$a_{(1)} > a_{(2)} > a_{(3)} > a_{(4)} > a_{(5)} > b_{(1)} > b_{(2)} > b_{(3)} > b_{(4)} > b_{(5)}$$

I would usually demonstrate the former by holding my two hands up with the two thumbs crossing each other (See Figure 0.5). The probability of rejecting H_0 when it is actually true is less than 0.01.

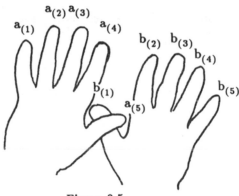

Figure 0.5

3) PERT.

Consider the so called Program Evaluation Review Technique (PERT). Suppose that there are n activities in a network representing a project and that there are three basic parameters describing the duration of the i-th activity. Let a_i, c_i and b_i denote the "optimistic duration", the "most probable duration", and the "pessimistic duration" respectively. The duration of the i-th activity is then usually assumed to follow a Beta distribution over (a_i, b_i) and to have mean duration m_i, where

$$m_i = \frac{a_i + 4c_i + b_i}{6}$$

and variance

$$\frac{(b_i - a_i)^2}{36}.$$

It is often argued that the probability distribution of the total duration of the whole project may be approximated by a Gaussian distribution, presumably by appealing to the Central Limit Theorem (CLT). Even ignoring the fact that the assumption of a Beta distribution is by no means indisputable, I would question the wisdom of a hasty use of the CLT. It is obvious that if a fair proportion of the n activities are in series, then a careful analysis is needed to check whether a Gaussian conclusion is tenable (See Chapter 7).

4) Experimental design.

Up to now, it seems to me that insufficient attention has been paid to non-linear designs. Past pre-occupation with linear models still seems to mask the important fact that these models are often unrealistic. I will not develop this any further (See Chapter 4).

5) Types of distributions.

It has been suggested that Pearson's Type III distribution be used to model the distribution of the waiting time for an "exceptionally big" flood (appropriately quantified). I would question the validity and wisdom of this approach in view of the scarcity of data which is inherent for this problem. It is even more unwise to expect that a sensible forecast for the next big flood may be obtainable from such a fitted distribution! The modelling and forecasting of such a phenomenon is probably more appropriately handled by a "point process" approach, in which the point events are epochs of flood. I understand that the problem is currently challenging the best brains in the field.

§ 0.9 Concluding Remarks.

If I were asked to say in a few words what I have learned during these last fifteen years of popularizing mathematical methods, I would without hesitation say that they have enabled me to appreciate the importance of the dictum

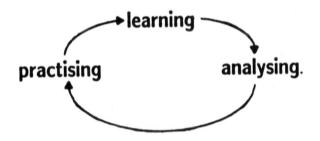

ON THE CALCULATION OF MINERAL RESERVES
AND
HILLSIDE AREAS ON CONTOUR MAPS*

HUA LOO-KENG AND WANG YUAN

§1. Introduction

Our geographers, mineralogists and geologists have presented many practical methods for calculating mineral reserves and hillside areas which enable us to engage in further investigations. In the present paper the authors attempt to compare these methods, to interpret their mutual relations and implicit errors, and to present some suggestions.

In mineral-body geometry (see [2]–[4]) one uses the Bauman formula, the frustum formula and the trapezoid formula for the approximate calculation of mineral reserves by divided layers. Suppose the mineral reserve volumes evaluated from these formulas are v, v_1 and v_2 respectively. In this paper we prove that they satisfy the inequality:

$$v \leqslant v_1 \leqslant v_2,$$

and we determine the conditions for taking the equal sign. The authors consider that the comparison of these three formulas should be based mainly on the number of dimensions, or degrees of freedom; consequently, we regard the Bauman formula as less limited than the others.

This paper presents a double-layer formula for the calculation of mineral reserves; this formula is due to our finding a new method for the proof of the Bauman formula. The proof is not only simple but also easy to improve still further. The advantages of our formula are in the consideration of more factors than in the Bauman formula without causing much complication; it also considers more factors than the Sobolevskiĭ formula (which generally uses a double layer formula for calculating the reserves, see [2]–[4]). We recommend it to technicians who engage in the calculation of mineral reserves.

For the calculation of hillside areas, the Volkov method (see [5]–[6]) is often used in geography and the Bauman method is often used in mineral-body geometry (see [1]–[2]). This paper points out that Bauman's method is more accurate than Volkov's, while both methods yield lower results than the exact value. Our paper completely determines those curved surfaces which can be handled with arbitrary accuracy by those two methods. In detail, the error depends on the change of the angle of inclination of the points on the surface. Only if the variation of the angle of

* Published in Acta Math. Sinica, **11** (1961) 1, 29–40. Translated from the Chinese by Hu Dihe.

25

inclination is small for all points on the surface can the Volkov method yield accurate results; and only if the angles of inclination of points between two adjacent contours differ from each other slightly can the Bauman method yield accurate results. Under other circumstances these two methods may give large errors. Therefore we suggest drawing several rays through the highest point on a contour map. If the surface is approximately a ruled surface, we may evaluate the individual surface areas between adjacent rays and then add them together. In case the variation of the angles of inclination for the surface between adjacent contours and adjacent rays is large, we may evaluate each individual surface area formed by rays and contours, and then add them together. The error of the results thus obtained will be relatively small.

§2. Calculation of mineral reserves

1. The Bauman method. Given a contour map of a mineral reserve with height difference h, the contour shown on the map represents essentially the *cross sectional area* at a fixed altitude. Let us evaluate the body volume between two such planes. The distance between these two planes is the height difference h. We denote the sections enclosed by the lower and the upper contours by A and B respectively (see figure 1; their areas are also represented by A and B). Bauman suggested evaluating the volume v between these two altitudes by

$$v = \left[\frac{1}{2}(A + B) - \frac{T(A,B)}{6} \right] h \tag{1}$$

where $T(A, B)$ is the area of the diagram shown below, and is called the Bauman correction number.

As shown in figure 2, draw the ray OP from the highest point O. The length of the section of the ray between A and B on the map is l. Construct another diagram, figure 3, take point O', then take $O'P' = l$ in the same direction as OP. When P moves one cycle around the contour, P' traces out a diagram whose area is called the Bauman correction number. Because of its dependence on sections A and B, we denote it by $T(A, B)$.

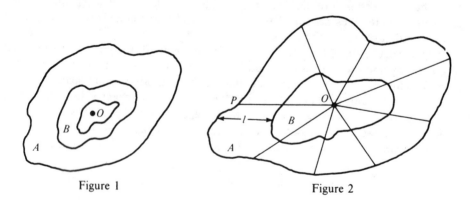

Figure 1   Figure 2

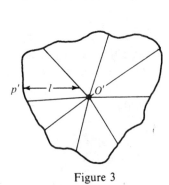

Figure 3

Figure 4

Adding up the computed volumes layer by layer, we obtain the mineral reserve volume V. In other words, supposing the areas enclosed by the $n + 1$ contours of the contour map to be $S_0, S_1, \ldots, S_n$, then the mineral reserve volume may be evaluated approximately by the formula

$$V = \left(\frac{S_0 + S_n}{2} + \sum_{m=1}^{n-1} S_m \right) h - \frac{h}{6} \sum_{m=0}^{n-1} T(S_m, S_{m+1}), \tag{2}$$

where h is contour distance (figure 4).

Theorem (BAUMAN). *Suppose that the lower surface A and the upper surface B (their areas are also represented by A and B) of a known body are planes, and A is parallel to B, h is the height between them, O is a point on B. If all the sections formed by the body and any arbitrary plane passing through O and perpendicular to B are quadrilateral, the body volume v is precisely as shown in formula (1).*

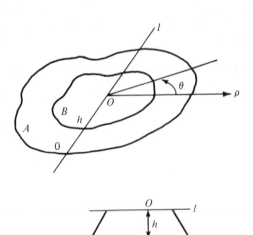

Figure 5

Proof. Introduce polar coordinates with origin O (see figure 5). Let the equation of the contour of altitude z in polar coordinates be

$$\rho = \rho(z,\theta), \qquad (0 \leqslant \theta \leqslant 2\pi),$$

where $\rho(z,0) = \rho(z,2\pi)$. Hereafter we will always assume that $\rho(z,\theta)$ $(0 \leqslant \theta \leqslant 2\pi$, $0 \leqslant z \leqslant h)$ is continuous. We may assume the altitudes of A and B to be 0 and h. Also let

$$\rho_1(\theta) = \rho(0,\theta), \qquad \rho_2(\theta) = \rho(h,\theta).$$

By assumption we have

$$\rho(Z,\theta) = \frac{Z}{h}\rho_2(\theta) + \frac{h-Z}{h}\rho_1(\theta) \qquad (0 \leqslant Z \leqslant h).$$

Hence the body volume is

$$\frac{1}{2}\int_0^h\int_0^{2\pi}\rho^2(Z,\theta)\,d\theta dZ = \frac{1}{2}\int_0^{2\pi}\int_0^h\left(\frac{Z}{h}\rho_2(\theta) + \frac{h-Z}{h}\rho_1(\theta)\right)^2 dZ\,d\theta$$

$$= \frac{h}{2}\int_0^{2\pi}\left(\frac{\rho_1^2(\theta)}{3} + \frac{\rho_2^2(\theta)}{3} + \frac{\rho_1(\theta)\rho_2(\theta)}{3}\right)d\theta$$

$$= \frac{h}{2}\left[\frac{1}{2}\int_0^{2\pi}\rho_1^2(\theta)\,d\theta + \frac{1}{2}\int_0^{2\pi}\rho_2^2(\theta)\,d\theta\right]$$

$$- \frac{h}{6}\left[\frac{1}{2}\int_0^{2\pi}(\rho_1(\theta) - \rho_2(\theta))^2 d\theta\right]$$

$$= \frac{h}{2}(A+B) - \frac{h}{6}T(A,B).$$

The theorem is proved.

2. *The relations between Bauman, frustum and trapezoid formulae.* If the bases of a body are in parallel planes, h is the altitude, O is a point on B, the following two formulae, in addition to the Bauman formula, are often used to evaluate the approximate volume of the body:

$$\text{Frustum formula:} \quad v_1 = \frac{h}{3}(A + B + \sqrt{AB}), \tag{3}$$

$$\text{Trapezoid formula:} \quad v_2 = \frac{h}{2}(A + B); \tag{4}$$

in general, it is appropriate to use formula (3) if $(A - B)/A > 40\%$, and formula (4) if $(A - B)/A < 40\%$.

Theorem 1. *The inequality*

$$v \leqslant v_1 \leqslant v_2 \tag{5}$$

holds for all cases. Also, $v = v_1$ if and only if the body is the frustum of a pyramid or a cone, and the perpendicular line from the vertex to the base passes through point O; while $v_1 = v_2$ if and only if $A = B$.

Proof. Use the same assumptions as in the Bauman theorem. From the Bauman formula and the Bunjakovskiĭ-Schwarz inequality, we have

$$v = \frac{h}{6} \int_0^{2\pi} \left(\rho_1^2(\theta) + \rho_2^2(\theta) + \rho_1(\theta)\rho_2(\theta) \right) d\theta$$

$$\leqslant \frac{h}{3} \left[\frac{1}{2} \int_0^{2\pi} \rho_1^2(\theta) \, d\theta + \frac{1}{2} \int_0^{2\pi} \rho_2^2(\theta) \, d\theta + \frac{1}{2} \sqrt{\int_0^{2\pi} \rho_1^2(\theta) \, d\theta \int_0^{2\pi} \rho_2^2(\theta) \, d\theta} \right]$$

$$= \frac{h}{3} \left[A + B + \sqrt{AB} \right] = v_1$$

if and only if $\rho_1(\theta) = C\rho_2(\theta)$ $(0 \leqslant \theta \leqslant 2\pi$, C is a constant); that is, $v = v_1$ if the body is a frustum and the perpendicular from the vertex to base A passes through the point O (figure 6).

Also, since

$$v_2 - v_1 = \frac{h}{2}(A + B) - \frac{h}{3}(A + B + \sqrt{AB}) = \frac{h}{6}\left(\sqrt{A} - \sqrt{B}\right)^2 \geqslant 0,$$

we have

$$v_1 \leqslant v_2,$$

with equality if and only if $A = B$. The theorem is proved.

These three formulae should be compared by looking at the number of dimensions. Since the dimension of a plane is 2, the formula obtained by considering this number (the number of degrees of freedom) to be 1 is less general.

Thus the trapezoid formula is obtained by regarding the middle section as the arithmetic mean of both bases, that is, regarding the number of degrees of freedom as 1.

The Bauman formula, however, regards the middle section to be of 2 degrees of freedom. To be precise, it assumes that $\rho(z, \theta)$ is obtained by a linear relationship of $\rho(0, \theta)$ and $\rho(h, \theta)$ (see 1).

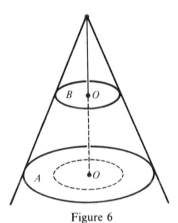

Figure 6

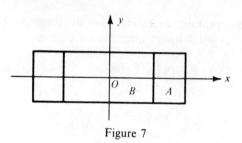

Figure 7

The frustum formula also regards the degrees of freedom of the middle section to be 2, but it further assumes that $\rho(0,\theta) = c\rho(h,\theta)$ $(0 \leqslant \theta \leqslant 2\pi)$, where c is a constant.

Therefore we think that the Bauman formula has more generality, and in general should give more accurate results for the approximate calculation of solid volumes. Nevertheless, we do not discard the possibility that the other two formulae may be more suitable for some specific bodies. For instance, consider a trapezoid with bases of equal width (as shown in figure 7). The trapezoid formula can yield its actual volume, while either the Bauman formula or the frustum formula will give a lower result. However, we should note that in this case the number of degrees of freedom of the section of the trapezoid is 1 (because the magnitude along the y-axis is constant).

We can also estimate the relative deviation of the trapezoid formula and the frustum formula with respect to Bauman formula.

For instance, when $(A - B)/A < 40\%$ (or $B > 3A/5$), the relative deviation of the result computed by the trapezoid formula and the result computed by the Bauman formula is given by

$$\Delta = \frac{v_2 - v}{v} = \frac{\frac{1}{2}(A+B)h - \frac{1}{2}(A+B)h + (h/6)T(A,B)}{\frac{1}{2}(A+B)h - (h/6)T(A,B)}$$

$$= \frac{T(A,B)}{3(A+B) - T(A,B)} .$$

Since

$$T(A,B) \leqslant A - B$$

(or $\frac{1}{2}\int_0^{2\pi}(\rho_1(\theta) - \rho_2(\theta))^2 d\theta \leqslant \frac{1}{2}\int_0^{2\pi}\rho_1^2(\theta)\, d\theta - \frac{1}{2}\int_0^{2\pi}\rho_2^2(\theta)\, d\theta$, an inequality which obviously holds), therefore

$$\Delta \leqslant \frac{A - B}{2A + 4B} .$$

By substituting the condition $B > \frac{3}{5}A$, we have

$$\Delta \leqslant \frac{A - \frac{3}{5}A}{2A + \frac{12}{5}A} = \frac{1}{11} < 10\%.$$

3. Suggestion of a formula for the calculation of mineral reserves. The Bauman formula is obtained by assuming the $\rho(z,\theta)$ to be a linear combination of $\rho(0,\theta)$ and $\rho(h,\theta)$. If we estimate two adjacent layers together, that is, if we know three adjacent contours $\rho(0,\theta)$, $\rho(h,\theta)$ and $\rho(zh,\theta)$, we obtain the approximation to the surface by joining the contours $\rho(0,\theta)$, $\rho(h,\theta)$ and $\rho(zh,\theta)$ with the surface constructed by parabolas; hence we suggest the following method of calculation.

Let A, B, C denote the sections enclosed by the three adjacent contours respectively (areas are also denoted by A, B, C), and let h be the distance between A and B as well as between B and C; then the total volume of these two layers can be approximately computed by the formula

$$v_3 = \frac{h}{3}(A + 4B + C) - \frac{h}{15}(2T(A,B) + 2T(B,C) - T(A,C)). \qquad (6)$$

If we omit the second term, equation (6) will be reduced identically to the familiar Sobolevskiĭ formula. Adding up the volumes of every double-layer, we obtain the approximate formula for the volume V of the total mineral reserve. In other words, supposing the areas enclosed by the $2n + 1$ contours of a contour map to be $S_0, S_1, \cdots, S_{2n}$, and the contour distance to be h, the mineral reserve volume V may be approximately computed by the formula

$$V = \frac{h}{3}\left[S_0 + S_{2n} + 4\sum_{i=0}^{n-1} S_{2i+1} + 2\sum_{i=1}^{n-1} S_{2i}\right]$$

$$- \frac{h}{15}\left[2\sum_{i=0}^{n-1} T(S_{2i}, S_{2i+1}) + 2\sum_{i=0}^{n-1} T(S_{2i+1}, S_{2i+2}) - \sum_{i=0}^{n-1} T(S_{2i}, S_{2i+2})\right]. \qquad (7)$$

Note. If the contour map contains an even number of contours, the uppermost layer can be estimated alone, and (7) can be used for the calculation of the rest of the layers.

Theorem 2. *Let the upper base C and the lower base A of a body be planes, let B be the middle section (areas are also denoted by C, A, B respectively), and A, C be parallel to B, let h be the distance between A and B and also between B and C, and let O be a point on C (figure 8). If the boundary of the section obtained in the plane*

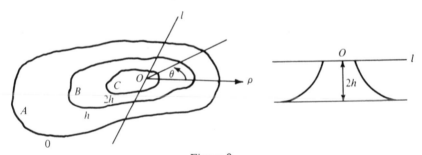

Figure 8

passing through O and perpendicular to C is composed of two straight lines and two parabolas, the volume v_3 of the body is precisely that shown in (6).

Proof. Introduce polar coordinates with origin O, let the equation of the contour of altitude z be

$$\rho = \rho(z,\theta) \quad (0 \leqslant \theta \leqslant 2\pi, \, \rho(z,0) = \rho(z,2\pi)).$$

We may also assume the altitudes of A, B, C to be 0, h, zh, and denote

$$\rho_1(\theta) = \rho(0,\theta), \, \rho_2(\theta) = \rho(h,\theta), \, \rho_3(\theta) = \rho(2h,\theta).$$

By assumption we have

$$\rho(z,\theta) = \frac{(z-h)(z-2h)}{2h^2}\rho_1(\theta) - \frac{z(z-2h)}{h^2}\rho_2(\theta) + \frac{z(z-h)}{2h^2}\rho_3(\theta). \quad (8)$$

Therefore the volume v_3 of the body is

$$\frac{1}{2}\int_0^{2h}\int_0^{2\pi}\rho^2(z,\theta)\,d\theta\,dz$$

$$= \frac{1}{2}\int_0^{2\pi}d\theta\int_0^{2h}\left[\frac{(z-h)(z-2h)}{2h^2}\rho_1(\theta) - \frac{z(z-2h)}{h^2}\rho_2(\theta) + \frac{z(z-h)}{2h^2}\rho_3(\theta)\right]^2 dz$$

$$= \frac{h}{2}\int_0^{2\pi}\left[\frac{4}{15}\rho_1^2(\theta) + \frac{16}{15}\rho_2^2(\theta) + \frac{4}{15}\rho_3^2(\theta) + \frac{4}{15}\rho_1(\theta)\rho_2(\theta)\right.$$

$$\left. + \frac{4}{15}\rho_2(\theta)\rho_3(\theta) - \frac{2}{15}\rho_1(\theta)\rho_3(\theta)\right]d\theta$$

$$= \frac{h}{2}\int_0^{2\pi}\left[\frac{\rho_1^2(\theta)}{3} + \frac{4\rho_2^2(\theta)}{3} + \frac{\rho_3^2(\theta)}{3} - \frac{2}{15}(\rho_1(\theta) - \rho_2(\theta))^2\right.$$

$$\left. - \frac{2}{15}(\rho_2(\theta) - \rho_3(\theta))^2 + \frac{1}{15}(\rho_1(\theta) - \rho_3(\theta))^2\right]d\theta$$

$$= \frac{h}{3}(A + 4B + C) - \frac{h}{15}(2T(A,B) + 2T(B,C) - T(A,C)).$$

The theorem is proved.

§3. Calculation of hillside areas

4. *The Bauman method and the Volkov method.* We first introduce methods often used by mineralogists and geographers. Suppose there is a map with contours of contour distance Δh; hereafter we will always assume there exists a summit and the contour is a closed curve (other conditions can easily be dealt with by extension from this case). Suppose that, starting from the highest point, we draw the contours (l_{n-1}), (l_{n-2}), $\cdots$, (l_0) one after the other (figure 9). Let the altitude of (l_0) be 0, and denote the altitude of (l_n) by h, the area between (l_i) and (l_{i+1}) by B_i (or the area of the projection).

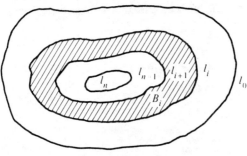

Figure 9

I. The procedure of the method frequently used in the mineral body geometry:

a. $C_i = \frac{1}{2}(l_i + l_{i+1})\Delta h$ (the area enclosed by the contour);

b. $\sum_{i=0}^{n-1}\sqrt{B_i^2 + C_i^2}$ is the asymptotic value of the hillside area (Bauman method).

II. The procedure of the method frequently used in geography:

a. $l = \sum_{i=0}^{n-1}l_i$ (the total length of the contours), $B = \sum_{i=0}^{n-1}B_i$ (the total area of the projections), $\operatorname{tg}\alpha = \Delta h \cdot l/B$ (the average angle of inclination);

b. $B\sec\alpha = \sqrt{B^2 + (\Delta h \cdot l)^2}$ is the asymptotic formula of the hillside area (Volkov method).

Note. $\sqrt{a^2 + b^2}$ can be obtained quickly by a graphical method (with the right-triangle formula).

Which one of these two methods is better? How close are the results obtained by these methods to the actual hillside area? In other words, as the number of the contours becomes infinite in such a way that $\Delta h \to 0$, what are the results given by these methods? Are they the actual hillside areas? In general, the answer is no. Only for some very special surfaces is the answer affirmative. Later we shall determine these specific surfaces and shall provide the relative differences between the results from these methods and the actual value, and at the same time indicate the procedures of computation to avoid large errors.

5. *Relation among Ba, Vo and S.* Introduce polar coordinates with the peak (l_n) as origin O. Let the equation of the contour of altitude z be

$$\rho = \rho(z,\theta) \quad (0 \leqslant \theta \leqslant 2\pi),$$

where $\rho(z,0) = \rho(z,\pi)$. Hereafter we always assume that

$$\frac{\partial\rho(z,\theta)}{\partial\theta} \quad \text{and} \quad \frac{\partial\rho(z,\theta)}{\partial z} \quad (0 \leqslant \theta \leqslant 2\pi, 0 \leqslant z \leqslant h)$$

are continuous. Let $z_i = h_i/n$; then the area enclosed by l_i is equal to

$$\frac{1}{2} \int_0^{2\pi} \rho^2(2_i, \theta) \, d\theta,$$

from which, by the mean-value theorem, we have

$$B_i = \frac{1}{2} \int_0^{2\pi} \left[\rho^2(z_i, \theta) - \rho^2(z_{i+1}, \theta) \right] d\theta$$

$$= - \int_0^{2\pi} \rho(z_i', \theta) \frac{\partial \rho(z_i', \theta)}{\partial z_i'} \, d\theta \, \Delta h,$$

where $z_i' \in [z_i, z_{i+1}]$, and $\Delta h = h/n$. The length of (l_i) is equal to

$$l_i = \int_{(l_i)} ds = \int_0^{2\pi} \sqrt{\rho^2(z_i, \theta) + \left(\frac{\partial \rho(z_i, \theta)}{\partial \theta} \right)^2} \, d\theta.$$

The result obtained from the Bauman method is

$$C_i = \int_0^{2\pi} \sqrt{\rho^2(z_i'', \theta) + \left(\frac{\partial \rho(z_i'', \theta)}{\partial \theta} \right)^2} \, d\theta \, \Delta h,$$

where the mean-value theorem has been applied with $z_i'' \in [z_i, z_{i+1}]$; hence as $\Delta h \to 0$, $\sum_{i=0}^{n-1} \sqrt{B_i^2 + C_i^2}$ approaches

$$\mathbf{Ba} = \int_0^h \sqrt{\left(\int_0^{2\pi} \rho \frac{\partial \rho}{\partial z} \, d\theta \right)^2 + \left(\int_0^{2\pi} \sqrt{\rho^2 + \left(\frac{\partial \rho}{\partial \theta} \right)^2} \, d\theta \right)^2} \, dz. \tag{9}$$

This is the value of the hillside area calculated by the Bauman method as $\Delta h \to 0$. It is also easy to see that

$$B = \frac{1}{2} \int_0^{2\pi} \rho^2(0, \theta) \, d\theta = \int_0^{2\pi} d\theta \int_0^h - \rho \frac{\partial \rho}{\partial z} \, dz$$

(note $\rho(h, \theta) = 0$) and the limit of $\Delta h \cdot l$ must be equal to

$$\lim_{n \to \infty} \sum_{i=0}^{n-1} \frac{h}{n} \int_0^{2\pi} \sqrt{\rho^2(z_i, \theta) + \left(\frac{\partial \rho(z_i, \theta)}{\partial \theta} \right)^2} \, d\theta$$

$$= \int_0^h dz \int_0^{2\pi} \sqrt{\rho^2 + \left(\frac{\partial \rho}{\partial \theta} \right)^2} \, d\theta;$$

therefore as $\Delta h \to 0$, the estimated value of the hillside area by the Volkov method approaches

$$\mathbf{Vo} = \sqrt{\left(\int_0^{2\pi} d\theta \int_0^h - \rho \frac{\partial \rho}{\partial z} \, dz \right)^2 + \left(\int_0^{2\pi} d\theta \int_0^h \sqrt{\rho^2 + \left(\frac{\partial \rho}{\partial \theta} \right)^2} \, dz \right)^2}. \tag{10}$$

Since

$$ds^2 = \left[\left(\frac{\partial\rho}{\partial\theta}\right)^2 + \rho^2\right]d\theta^2 + 2\frac{\partial\rho}{\partial\theta}\frac{\partial\rho}{\partial z}d\theta\,dz + \left(1 + \left(\frac{\partial\rho}{\partial z}\right)^2\right)dz^2,$$

the area S of the hillside surface is

$$S = \int_0^{2\pi} d\theta \int_0^h \sqrt{\rho^2 + \left(\frac{\partial\rho}{\partial\theta}\right)^2 + \left(-\rho\frac{\partial\rho}{\partial z}\right)^2}\,d\theta. \tag{11}$$

For the comparison of **Ba**, **Vo** and S, we introduce a function of a complex variable

$$f(z,\theta) = \rho\frac{\partial\rho}{\partial z} + i\sqrt{\rho^2 + \left(\frac{\partial\rho}{\partial\theta}\right)^2}, \tag{12}$$

and thus obtain

$$\mathbf{Ba} = \int_0^h \left|\int_0^{2\pi} f(z,\theta)\,d\theta\right| dz, \tag{13}$$

$$\mathbf{Vo} = \left|\int_0^h \int_0^{2\pi} f(z,\theta)\,d\theta\,dz\right|, \tag{14}$$

and

$$S = \int_0^h \int_0^{2\pi} |f(z,\theta)|\,d\theta\,dz. \tag{15}$$

Hence the inequality

$$\mathbf{Vo} \leqslant \mathbf{Ba} \leqslant S \tag{16}$$

is obviously established.

Therefore we recognize: (i) the Bauman method is more accurate than the Volkov method; (ii) the obtained results are smaller than the actual value; (iii) because the Bauman method yields a smaller result, we can make the modification $C_i = l_i\Delta h$. Thus we not only simplify the computation but also increase the value.

Before investigating the surface for which $\mathbf{Vo} = S$ and $\mathbf{Ba} = S$, we introduce the following lemma:

Lemma. *If $f(x)$ is a complex function in the interval $[a,b]$, where a and b are real numbers, the necessary and sufficient condition for the equality*

$$\left|\int_a^b f(x)\,dx\right| = \int_a^b |f(x)|\,dx \tag{17}$$

is that the ratio of the imaginary part to the real part of $f(x)$ should be constant.

Proof. Let $f(x) = \rho(x)e^{i\theta(x)}$, $\rho(x) \geqslant 0$ and $\theta(x)$ is a real function. It is obvious that (17) is established provided that $\theta(x)$ is a constant with respect to x. Conversely, since

$$\left(\left|\int_a^b f(x)\,dx\right|\right)^2 = \int_a^b \int_a^b f(x)\,\overline{f(y)}\,dx\,dy = \int_a^b \int_a^b \rho(x)\rho(y)e^{i(\theta(x)-\theta(y))}dx\,dy$$

$$= 2 \iint_{a\leqslant x < y \leqslant b} \rho(x)\rho(y)\cos\left[\theta(x)-\theta(y)\right]dx\,dy,$$

$$\left(\int_a^b |f(x)|\,dx\right)^2 = 2 \iint_{a\leqslant x < y \leqslant b} \rho(x)\rho(y)\,dx\,dy.$$

Therefore if (17) is established, there must exist

$$\cos(\theta(x)-\theta(y)) \equiv 1,$$

or $\theta(x) \equiv \theta(y)$, as the lemma asserts.

It is easy to see that the lemma holds for multiple integrals, too.

From the lemma we see that the necessary and sufficient condition for the establishment of

$$\mathbf{Vo} = \left|\int_0^{2\pi}\int_0^h f(z,\theta)\,dz\,d\theta\right| = \int_0^{2\pi}\int_0^h |f(z,\theta)|\,dz\,d\theta = S$$

is that the ratio of the imaginary part to the real part of $f(z,\theta)$ should be a constant c; hence we obtain the partial differential equation

$$\rho^2 + \left[\frac{\partial\rho}{\partial\theta}\right]^2 = c^2\left[\rho\frac{\partial\rho}{\partial z}\right]^2. \tag{18}$$

In other words, only for those functions $\rho = \rho(z,\theta)$ which satisfy this partial differential equation can the Volkov method yield an exact answer. These, of course, should meet the following conditions: $\rho(h,\theta) = 0$ and $\rho(0,\theta) = \rho_0(\theta)$ (the equation of the base of the surface).

We will not solve this partial differential equation; instead, we will try to see its geometric meaning. Regard θ and z as parameters, that is,

$$x = \rho\cos\theta, \qquad y = \rho\sin\theta, \qquad z = z,$$

and ρ as a function of θ and z. From

$$\frac{\partial x}{\partial\theta} = \frac{\partial\rho}{\partial\theta}\cos\theta - \rho\sin\theta, \qquad \frac{\partial y}{\partial\theta} = \frac{\partial\rho}{\partial\theta}\sin\theta + \rho\cos\theta, \qquad \frac{\partial z}{\partial\theta} = 0,$$

$$\frac{\partial x}{\partial z} = \frac{\partial\rho}{\partial z}\cos\theta, \qquad \frac{\partial y}{\partial z} = \frac{\partial\rho}{\partial z}\sin\theta, \qquad \frac{\partial z}{\partial z} = 1$$

we know that the direction of the normal at the point (θ, z) of the surface is

$$\left(\frac{\partial\rho}{\partial\theta}\sin\theta + \rho\cos\theta, \; -\frac{\partial\rho}{\partial\theta}\cos\theta + \rho\sin\theta, \; -\rho\frac{\partial\rho}{\partial z}\right).$$

From (18) we see that the cosine of the angle α between the normal and the z-axis (or the inclination at point (θ, z)) is equal to

$$\cos\alpha = \frac{-\rho(\partial\rho/\partial z)}{\sqrt{(\rho\partial\rho/\partial z)^2 + (\partial\rho/\partial\theta)^2 + \rho^2}} = \frac{1}{\sqrt{1+c^2}},$$

which is a constant. That is to say, the tangent plane meets the horizontal plane (xy-plane) at a fixed angle α. We will now interpret the geometric properties of such surfaces.

Construct an arbitrary vertical plane from the high-water mark to the xy-plane. Then tangent line of the curve at every point has the same intersection angle with the xy-plane. Hence it is a straight line.

Construct a base from any closed plane curve (l_0). Take any point (l_n) with projection inside the base as the high-water mark. The straight line passing through the high-water mark and perpendicular to the base is called the axis. Through any point A in (l_0), construct a straight line which lies in the plane formed by A and the axis, and meets the base at an angle α. The diagram constructed by such straight lines is the one which satisfies $\mathbf{Vo} = S$.

Therefore, if there exists a peak and no steep angle downward, then these will be only surfaces whose bases are circles, or polygons formed by tangent lines of a circle, or a construction of some circular arcs and some tangent lines, and whose axis is the line passing through the center of the circle and perpendicular to the base (see figure 10).

Among well-known examples, only the Mongolian tent, the pyramid, and diagrams combined from these can be arbitrarily well approximated by the Volkov method.

But when does $\mathbf{Ba} = S$? Of course, when $\mathbf{Vo} = S$, $\mathbf{Ba} = S$. Is there any surface other than those mentioned above? Yes, we demonstrate it as follows: from

$$\mathbf{Ba} = \int_0^h \left| \int_0^{2\pi} f(z,\theta)\, d\theta \right| dz = \int_0^h \int_0^{2\pi} |f(z,\theta)|\, d\theta\, dz = S$$

we obtain

$$\int_0^h \left(\int_0^{2\pi} |f(z,\theta)|\, d\theta - \left| \int_0^{2\pi} f(z,\theta)\, d\theta \right| \right) dz = 0.$$

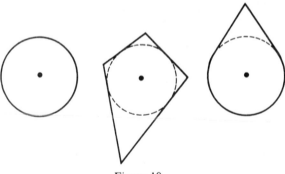

Figure 10

Since the integrand is not negative, we have for any z that

$$\int_0^{2\pi} |f(z,\theta)|\, d\theta = \left| \int_0^{2\pi} f(z,\theta)\, d\theta \right|.$$

Therefore, when we fix z, the ratio of the imaginary part to the real part of $f(z,\theta)$ is constant, that is, c in equation (18) is a function of z alone. Hence, only those surfaces with equal inclination for the same altitude can satisfy $\mathbf{Ba} = S$. Among well-known examples, only the Gourd and the White Pagoda (in the North Sea) can be approximated arbitrarily well by the Bauman method.

Now we estimate the error for these two methods. Suppose that the cosines of inclination of the points on the surface are between two positive constants ξ and η,

$$\xi \leqslant \cos \alpha \leqslant \eta,$$

or

$$\xi \leqslant \frac{-\rho(\partial\rho/\partial z)}{\sqrt{(\rho(\partial\rho/\partial z))^2 + (\partial\rho/\partial\theta)^2 + \rho^2}} \leqslant \eta.$$

From this we obtain

$$\frac{\rho^2 + (\partial\rho/\partial\theta)^2}{(\rho(\partial\rho/\partial z))^2 + (\partial\rho/\partial\theta)^2 + \rho^2} \geqslant 1 - \eta^2,$$

so that

$$\int_0^{2\pi} \int_0^h \sqrt{\rho^2 + (\partial\rho/\partial\theta)^2}\, dz\, d\theta$$

$$\geqslant \sqrt{1 - \eta^2} \int_0^{2\pi} d\theta \int_0^h \sqrt{(\rho(\partial\rho/\partial z))^2 + (\partial\rho/\partial\theta)^2 + \rho^2}\, dz = \sqrt{1 - \eta^2}\, S,$$

$$\int_0^{2\pi} \int_0^h -\rho \frac{\partial\rho}{\partial z}\, dz\, d\theta \geqslant \xi S,$$

from which

$$\mathbf{Vo} \geqslant \sqrt{\xi^2 S^2 + (1 - \eta^2) S^2} = \sqrt{1 + \xi^2 - \eta^2}\, S.$$

Also because $1 > \eta \geqslant \xi > 0$, hence

$$\frac{\xi}{\eta} \leqslant \sqrt{1 + \xi^2 - \eta^2},$$

(squaring both sides of this inequality, we have $(\eta^2 - \xi^2)(1 - \eta^2) \geqslant 0$) and we obtain

$$\mathbf{Vo} \geqslant \frac{\xi}{\eta} S.$$

Thus we have proved the following theorem.

Theorem 3. *If all the cosines of the inclinations α at any point on the surface $\rho = \rho(z,\theta)$ $(0 \leqslant z \leqslant h, 0 \leqslant \theta \leqslant 2\pi)$ satisfy $0 < \xi \leqslant \cos\alpha \leqslant \eta$, then the inequality*

$$\frac{\xi}{\eta} S \leqslant Vo \leqslant \mathbf{Ba} \leqslant S \tag{19}$$

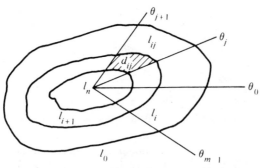

Figure 11

is established. The necessary and sufficient condition for $\mathbf{Vo} = S$ *is that any point on the surface has equal inclination; the necessary and sufficient condition for* $\mathbf{Ba} = S$ *is that the points at equal altitude have the same inclination.*

6. *Suggestions for methods of calculation.* From theorem 3 we can see that the Volkov method can give accurate results only when the variation of the inclination of points on the surface is small, and the Bauman method can give accurate results only when the difference of inclination of points between two adjacent altitudes is small. For other cases these methods may yield greater error.

Accordingly, we suggest the following methods of calculation: introduce several rays $\theta_0, \theta_1, \cdots, \theta_{m-1}$ from the high-water mark (l_n) on the contour map (see figure 11), where the amplitude of θ_j is equal to $2\pi j/m$. The area enclosed by rays θ_j, θ_{j+1} and contours l_i, l_{i+1} is denoted by d_{ij}; the length of the section of l_i intercepted by θ_j and θ_{j+1} is denoted by l_{ij}.

Method I.

a. $D_j = \sum_{i=0}^{n-1} d_{ij}$ (the area on the contour map between rays θ_j and θ_{j+1});
b. $E_j = [\sum_{i=0}^{n-1} l_{ij}]\Delta h$ (the sum of the areas of the vertical screen between two vertical walls);
c. $\sigma_1 = \sum_{j=0}^{m-1} \sqrt{D_j^2 + E_j^2}$ is the asymptotic value of the surface area.

Method II.

a. $e_{ij} = l_{ij}\Delta h$ (the area of the vertical screen between two vertical walls);
b. $\sigma_2 = \sum_{i=0}^{n-1}\sum_{j=0}^{m-1} \sqrt{d_{ij}^2 + e_{ij}^2}$ is the asymptotic value of the surface area.

In the same manner as in the preceding section we know that

$$K = \int_0^{2\pi} \sqrt{\left(\int_0^h -\rho\frac{\partial\rho}{\partial z}\,dz\right)^2 + \left(\int_0^h \sqrt{\rho^2 + (\partial\rho/\partial\theta)^2}\,dz\right)^2}\,d\theta$$

$$= \int_0^{2\pi} \left|\int_0^h f(z,\theta)\,dz\right|\,d\theta \tag{20}$$

and

$$S = \int_0^{2\pi} \int_0^h \sqrt{\rho^2 + (\partial\rho/\partial\theta)^2 + (\rho(\partial\rho/\partial z))^2} \, dz \, d\theta$$

$$= \int_0^{2\pi} \int_0^h |f(z,\theta)| \, dz \, d\theta \tag{21}$$

are the values which σ_1 and σ_2 approach as $n \to \infty$, $m \to \infty$ (see (12) for the definition of $f(z,\theta)$).

Evidently, $\mathrm{Vo} \leqslant K \leqslant S$ (see (10)), and in the same manner as in the preceding section we see that the necessary and sufficient condition for $K = S$ is that the surface is a ruled surface. Since σ_2 approaches the actual area, method II is the most accurate and reliable.

REFERENCES

[1] V. I. Bauman, On the calculation of mineral reserves, *Gornyĭ žurnal*, December, 1908. (Russian)

[2] I. N. Ušakov, *Mineral reserve geometry*, Coal Industry Publishing House, 1957. (Chinese)

[3] P. A. Ryžov, *Mineral body geometry*, Geology Publishing House, 1957. (Chinese)

[4] S. S. Izakson, *The verification of calculation and the determination of calculation error of mineral reserves*, Coal Industry Publishing House, 1958. (Chinese)

[5] N. M. Volkov, *Principles and methods of the measurement of a diagram*, 1950. (Chinese)

[6] Lu Shu-fen, "Problems of the calculation of ground surface area on a contour map," *Journal of Mensuration* **4** (1960), no. 1. (Chinese)

CHAPTER 2

THE MESHING GEAR-PAIR PROBLEM

§ 2.1 Introduction.

The meshing gear-pair problem (Gua Lun Wen Ti in Chinese) was raised by Zhuang Yi Ming, an ordinary worker in China. For the design of two gear-pairs he wanted to know optimal gear ratios with the numbers of teeth chosen from the integers 20, 21, ... , 100 so that the output speed is as close to π as possible. This is equivalent to finding four integers a, b, c, d between 20 and 100 such that

$$\left| \pi - \frac{ab}{cd} \right|$$

is a minimum. This is a problem of diophantine approximation. A general form of the problem is the following.

<u>Problem.</u> Given a real number α, integers $k \leq K$ and $l \leq L$, and two sets of positive integers $X = \{x_1, \dots, x_K\}$ and $Y = \{y_1, \dots, y_L\}$, find k integers $x_{i_1}, \dots, x_{i_k}$ of X and l integers $y_{j_1}, \dots, y_{j_l}$ of Y such that

$$\left| \alpha - \frac{x_{i_1} \cdots x_{i_k}}{y_{j_1} \cdots y_{j_l}} \right|$$

is a minimum.

In this chapter we will provide an algorithm for solving the above problem that uses continued fractions and Farey means. As an example we will give the solution to Zhuang's original problem.

§ 2.2 Simple continued fractions.

Any real non-integer number α may be written as

$$\alpha = a_0 + \frac{1}{\alpha_1}, \quad a_0 = \lfloor \alpha \rfloor, \quad \alpha_1 > 1,$$

where $\lfloor \alpha \rfloor$ denotes the integral part of α. Further, if α_1 is not an integer, let

$$\alpha_1 = a_1 + \frac{1}{\alpha_2}, \quad a_1 = \lfloor \alpha_1 \rfloor, \quad \alpha_2 > 1,$$

and in general, if α_{n-1} is not an integer, let

$$\alpha_{n-1} = a_{n-1} + \frac{1}{\alpha_n}, \quad a_{n-1} = \lfloor \alpha_{n-1} \rfloor, \quad \alpha_n > 1.$$

Then we have an expansion of α :

$$\alpha = a_0 + \cfrac{1}{a_1 + \cfrac{1}{a_2 + \cfrac{}{\ddots + \cfrac{1}{a_n + \cfrac{1}{\ddots}}}}}$$

where each a_i $(i \geq 1)$ is a positive integer. Shorthand notations are

$$\alpha = a_0 + \frac{1}{a_1} + \frac{1}{a_2} + \cdots \frac{1}{a_n} + \cdots$$

and

$$\alpha = [a_0, a_1, a_2, \ldots, a_n, \ldots]. \tag{2.1}$$

This is called the expansion of α into the simple continued fraction. It is well known that if the above expansion is finite, then α is rational, otherwise α is irrational (see e.g. Hua [1982], p. 251; Hardy and Wright [1960], pp. 135-136). Since

$$[a_0, a_1, a_2, \ldots, a_n, 1] = [a_0, a_1, a_2, \ldots, a_n + 1],$$

the expansion of α into a simple continued fraction is unique if the last number in its expansion is required to be larger than 1. The expansion of any irrational α is always unique.

Given that (2.1) is the simple continued fraction of α, the finite simple continued fraction

$$[a_0, a_1, a_2, \ldots, a_n]$$

is called the n-<u>th</u> <u>convergent</u> of α.

<u>Theorem</u> 2.1. Let p_n and q_n be defined by

$$p_0 = a_0, \quad p_1 = a_1 a_0 + 1, \quad p_n = a_n p_{n-1} + p_{n-2} \quad (n \geq 2)$$

$$q_0 = 1, \quad q_1 = a_1, \quad q_n = a_n q_{n-1} + q_{n-2} \quad (n \geq 2).$$

Then $\dfrac{p_n}{q_n}$ is the n-th convergent of α.

Proof. In the expansion of α we have that a_i is a positive integer for all $i \geq 1$. We will prove the theorem for all simple continued fractions with $a_i > 0$ and real $(i \geq 1)$.

Since

$$[a_0] = \frac{a_0}{1}, \quad [a_0, a_1] = \frac{a_1 a_0 + 1}{a_1}, \quad [a_0, a_1, a_2] = \frac{a_2(a_1 a_0 + 1) + a_0}{a_2 a_1 + 1},$$

the theorem holds for $n = 0, 1$ and 2. Now suppose that $m \geq 2$ and that the theorem holds for $0 \leq n \leq m$. We have

$$[a_0, a_1, \ldots, a_{m-1}, a_m, a_{m+1}] = [a_0, a_1, \ldots, a_{m-1}, a_m + \frac{1}{a_{m+1}}],$$

and, using the induction assumption, these two simple continued fractions have identical p_n's and q_n's for all $0 \leq n \leq m-1$. By applying the induction assumption for $n = m$ to the expression on the right, we obtain

$$[a_0, a_1, \ldots, a_{m-1}, a_m, a_{m+1}] = \frac{(a_m + \frac{1}{a_{m+1}})p_{m-1} + p_{m-2}}{(a_m + \frac{1}{a_{m+1}})q_{m-1} + q_{m-2}}$$

$$= \frac{a_{m+1}(a_m p_{m-1} + p_{m-2}) + p_{m-1}}{a_{m+1}(a_m q_{m-1} + q_{m-2}) + q_{m-1}}$$

$$= \frac{a_{m+1}p_m + p_{m-1}}{a_{m+1}q_m + q_{m-1}} = \frac{p_{m+1}}{q_{m+1}}.$$

The theorem follows by induction. □

Theorem 2.2. We have

$$p_n q_{n-1} - p_{n-1}q_n = (-1)^{n-1} \quad (n \geq 1) \qquad (2.2)$$

and

$$p_n q_{n-2} - p_{n-2}q_n = (-1)^n a_n \quad (n \geq 2).$$

Proof. Clearly (2.2) is true for $n = 1$. Now suppose that (2.2) is true for all positive integers less than n. Then, from Theorem 2.1, we have

$$p_n q_{n-1} - p_{n-1}q_n =$$

$$= (a_n P_{n-1} + P_{n-2})q_{n-1} - P_{n-1}(a_n q_{n-1} + q_{n-2})$$

$$= P_{n-2}q_{n-1} - P_{n-1}q_{n-2} = (-1)^{n-1}.$$

Hence (2.2) follows by induction.

From Theorem 2.1 and (2.2), we have

$$P_n q_{n-2} - P_{n-2}q_n =$$

$$= (a_n P_{n-1} + P_{n-2})q_{n-2} - P_{n-2}(a_n q_{n-1} + q_{n-2})$$

$$= a_n(P_{n-1}q_{n-2} - P_{n-2}q_{n-1}) = (-1)^n a_n.$$

The theorem is proved. □

Let $\alpha_n = [a_n, a_{n+1}, \dots]$. α_n is called the $n+1$-th complete quotient of the real number

$$\alpha = [a_0, a_1, \dots, a_n, \dots].$$

Theorem 2.3. We have

$$\alpha = \alpha_0, \quad \alpha_1 = \frac{\alpha_1 a_0 + 1}{1}, \quad \alpha = \frac{\alpha_n P_{n-1} + P_{n-2}}{\alpha_n q_{n-1} + q_{n-2}} \quad (n \geq 2).$$

Proof. Since

$$\alpha = \alpha_0, \quad \alpha = a_0 + \frac{1}{\alpha_1} = \frac{\alpha_1 a_0 + 1}{\alpha_1},$$

$$\alpha = a_0 + \frac{1}{a_1 + \frac{1}{\alpha_2}} = \frac{\alpha_2(a_1 a_0 + 1) + a_0}{\alpha_2 a_1 + 1} = \frac{\alpha_2 P_1 + P_0}{\alpha_2 q_1 + q_0},$$

the theorem is true for $n \leq 2$. Now let $n > 2$ and suppose the theorem is true for all $m < n$. Then

$$\alpha = \frac{\alpha_{n-1} p_{n-2} + p_{n-3}}{\alpha_{n-1} q_{n-2} + q_{n-3}} = \frac{(a_{n-1} + \frac{1}{\alpha_n}) p_{n-2} + p_{n-3}}{(a_{n-1} + \frac{1}{\alpha_n}) q_{n-2} + q_{n-3}}$$

$$= \frac{\alpha_n(a_{n-1} p_{n-2} + p_{n-3}) + p_{n-2}}{\alpha_n(a_{n-1} q_{n-2} + q_{n-3}) + q_{n-2}} = \frac{\alpha_n p_{n-1} + p_{n-2}}{\alpha_n q_{n-1} + q_{n-2}},$$

and the theorem is proved. $\square$

Theorem 2.4. We have

$$\alpha - \frac{p_n}{q_n} = \frac{(-1)^n}{q_n(\alpha_{n+1} q_n + q_{n-1})} \qquad (n \geq 0),$$

with the understanding that $q_{-1} = 0$.

Note the theorem holds only for $0 \leq n < m$ if $\alpha = [a_0, a_1, \ldots, a_m]$.

Proof. From Theorem 2.3 and the convention $p_{-1} = 1$, we have for $n \geq 0$

$$\alpha - \frac{p_n}{q_n} = \frac{\alpha_{n+1} p_n + p_{n-1}}{\alpha_{n+1} q_n + q_{n-1}} - \frac{p_n}{q_n}$$

$$= \frac{-(p_n q_{n-1} - p_{n-1} q_n)}{q_n(\alpha_{n+1} q_n + q_{n-1})} = \frac{(-1)^n}{q_n(\alpha_{n+1} q_n + q_{n-1})}.$$

The theorem is proved. $\square$

Theorem 2.5. We have

$$\left| \alpha - \frac{p_n}{q_n} \right| \leq \frac{1}{q_n q_{n+1}}$$

and

$$|q_n \alpha - p_n| < |q_{n-1}\alpha - p_{n-1}|.$$

Proof. It follows from Theorem 2.4 that

$$\left| \alpha - \frac{p_n}{q_n} \right| = \frac{1}{q_n(\alpha_{n+1}q_n + q_{n-1})}$$

$$\leq \frac{1}{q_n(a_{n+1}q_n + q_{n-1})} = \frac{1}{q_n q_{n+1}},$$

and from Theorems 2.1 and 2.4 it follows that

$$|q_{n-1}\alpha - p_{n-1}| = \frac{1}{\alpha_n q_{n-1} + q_{n-2}} > \frac{1}{(a_n + 1)q_{n-1} + q_{n-2}}$$

$$= \frac{1}{q_n + q_{n-1}} > \frac{1}{\alpha_{n+1}q_n + q_{n-1}} = |q_n \alpha - p_n|.$$

The theorem is proved. □

§ 2.3 Farey series.

The Farey series $\mathfrak{F}_n$ of order n is the ascending series of irreducible fractions between 0 and 1 whose denominators do not exceed n. Thus $\frac{a}{b}$ belongs to $\mathfrak{F}_n$ if

$$0 \leq a \leq b \leq n \quad \text{and} \quad (a, b) = 1,$$

where (a, b) denotes the greatest common divisor of a and b. If $\frac{a}{b}$ and $\frac{a'}{b'}$ are two successive terms in the Farey series $\mathfrak{F}_n$, then

$$\frac{a + a'}{b + b'}$$

is called their <u>Farey</u> mean.

Theorem 2.6. Suppose that a, a', b and b' are four positive integers such that

$$a'b - ab' = 1.$$

Then there is no fraction between $\frac{a}{b}$ and $\frac{a'}{b'}$ with denominator smaller than $b + b'$ or with numerator smaller than $a + a'$.

Proof. Since $\frac{a}{b}$ and $\frac{a'}{b'}$ can be changed to $\frac{b'}{a'}$ and $\frac{b}{a}$ respectively, it is sufficient to prove that there is no fraction between $\frac{a}{b}$ and $\frac{a'}{b'}$ with denominator smaller than $b + b'$.

Clearly

$$\frac{a}{b} < \frac{a + a'}{b + b'} < \frac{a'}{b'}$$

and

$$(a + a')b - a(b + b') = a'(b + b') - (a + a')b' = a'b - ab' = 1,$$

and therefore $(a + a', b + b') = 1$. Every fraction α in the interval $(\frac{a}{b}, \frac{a'}{b'})$ other than $\frac{a + a'}{b + b'}$ must satisfy one of the two inequalities

$$\frac{a}{b} < \alpha < \frac{a + a'}{b + b'} \quad \text{and} \quad \frac{a + a'}{b + b'} < \alpha < \frac{a'}{b'}.$$

It is therefore sufficient to prove that there is no fraction between $\frac{a}{b}$ and $\frac{a'}{b'}$ with denominator smaller than $\text{MAX}\{b, b'\}$.

Suppose that a''/b'' satisfies

$$\frac{a}{b} < \frac{a''}{b''} < \frac{a'}{b'} \quad \text{and} \quad b'' < \text{MAX}\{b, b'\}.$$

If $b' \geq b$, then $b'' < b'$ and

$$\frac{a''}{b''} - \frac{a}{b} = \frac{a''b - ab''}{bb''} \geq \frac{1}{bb''} > \frac{1}{bb'} = \frac{a'}{b'} - \frac{a}{b},$$

which leads to a contradiction. If $b' < b$, then $b'' < b$ and

$$\frac{a'}{b'} - \frac{a''}{b''} = \frac{a'b'' - a''b'}{b'b''} \geq \frac{1}{b'b''} > \frac{1}{bb'} = \frac{a'}{b'} - \frac{a}{b},$$

and so we also have a contradiction. The theorem is proved. □

It follows immediately that if $\frac{a'}{b'} \leq 1$, then $\frac{a}{b}$ and $\frac{a'}{b'}$ are successive terms in the Farey series $\mathcal{F}_{b+b'-1}$ and

$$\frac{a+a'}{b+b'}$$

is their Farey mean.

§ 2.4 An algorithm for the problem.

In this section we will give an algorithm that finds the solution of the problem mentioned in section 2.1. Assume that

$$\frac{(\text{MIN } X)^k}{(\text{MAX } Y)^l} < \alpha < \frac{(\text{MAX } X)^k}{(\text{MIN } Y)^l},$$

otherwise the solution to the problem is obvious.

1. Expand α into the simple continued fraction

$$\alpha = [\, a_0, a_1, \ldots, a_n, \ldots \,].$$

Then it follows from Theorems 2.2, 2.4 and 2.5 that the convergents to α satisfy

$$\frac{p_0}{q_0} < \frac{p_2}{q_2} < \cdots < \frac{p_{2m}}{q_{2m}} < \cdots < \alpha < \cdots < \frac{p_{2m-1}}{q_{2m-1}} < \cdots < \frac{p_3}{q_3} < \frac{p_1}{q_1} \quad (2.3)$$

and

$$\cdots < \left| \alpha - \frac{p_m}{q_m} \right| < \left| \alpha - \frac{p_{m-1}}{q_{m-1}} \right| < \cdots < \left| \alpha - \frac{p_0}{q_0} \right|. \quad (2.4)$$

Let

$$x = \underset{i}{\text{MAX}} \; x_i, \quad y = \underset{j}{\text{MAX}} \; y_j.$$

Let n be the largest integer such that

$$p_n \le x^k \quad \text{and} \quad q_n \le y^l$$

hold simultaneously. Then, from Theorems 2.2 and 2.6, there is no fraction between

$$\frac{p_n}{q_n} \quad \text{and} \quad \frac{p_{n+1}}{q_{n+1}}$$

with denominator $\le y^l$ and with numerator $\le x^k$, i.e. there is no fraction of the required form in this interval. Therefore we should start by trying to find the required fraction between

$$\frac{p_{n-1}}{q_{n-1}} \quad \text{and} \quad \frac{p_n}{q_n}.$$

2. We may suppose without loss of generality that n is odd. Then

$$\frac{p_{n-1}}{q_{n-1}} < \alpha < \frac{p_n}{q_n}.$$

It follows from Theorem 2.6 that all fractions in the interval $[\frac{p_{n-1}}{q_{n-1}}, \frac{p_n}{q_n}]$ with denominator $\le y^l$ and numerator $\le x^k$ can be found among the Farey means

$$\frac{p_{n-1} + p_n}{q_{n-1} + q_n} , \quad \frac{2p_{n-1} + p_n}{2q_{n-1} + q_n} , \quad \frac{p_{n-1} + 2p_n}{q_{n-1} + 2q_n} , \dots .$$

Now we determine which of these Farey means can be expressed in the form

$$\frac{x_{i_1} \dots x_{i_k}}{y_{j_1} \dots y_{j_l}} .$$

Suppose such fractions do exist and among them $\frac{a}{b}$ is the closest to α. If

$$\left| \frac{a}{b} - \alpha \right| \leq \left| \frac{p_n}{q_n} - \alpha \right| , \tag{2.5}$$

then it follows from (2.3) and (2.4) that $\frac{a}{b}$ is the solution to our problem. On the other hand, if none of the above fractions can be expressed in the desired form, or if (2.5) does not hold, then the search will be continued among appropriate Farey means in the intervals

$$[\frac{p_{n-1}}{q_{n-1}}, \frac{p_{n-2}}{q_{n-2}}], \quad [\frac{p_{n-3}}{q_{n-3}}, \frac{p_{n-2}}{q_{n-2}}], \quad [\frac{p_{n-3}}{q_{n-3}}, \frac{p_{n-4}}{q_{n-4}}], \dots$$

and so on, until the solution is found.

§ 2.5 The solution to the meshing gear-pair problem.

Problem. Find four integers a, b, c, d between 20 and 100 such that

$$\left| \pi - \frac{ab}{cd} \right|$$

is a minimum.

Solution. 1. The expansion of π into the simple continued fraction is

$$\pi = [\,3,\, 7,\, 15,\, 1,\, 292,\, \ldots\,].$$

Its convergents are

$$\frac{3}{1},\quad \frac{22}{7},\quad \frac{333}{106},\quad \frac{355}{113},\quad \frac{103993}{33102},\quad \ldots\ .$$

We will use the notation of section 2.4. Since $k = l = 2$, $x^2 = y^2 = 10{,}000$ and $n = 3$, we will start to search for fractions of the form $\frac{a\,b}{c\,d}$ in the interval $[\frac{333}{106}, \frac{355}{113}]$.

2. We determine the Farey means of the form

$$\frac{333 + (k \times 355)}{106 + (k \times 113)}$$

with both numerator and denominator $\leq 10{,}000$. They are

$$\frac{333}{106} < \frac{688}{219} < \frac{1043}{332} < \frac{1398}{445} < \frac{1753}{558} < \frac{2108}{671} < \frac{2463}{784} < \frac{2818}{897}$$

$$< \frac{3173}{1010} < \frac{3528}{1123} < \frac{3883}{1236} < \frac{4238}{1349} < \frac{4593}{1462} < \frac{4948}{1575} < \frac{5303}{1688} < \frac{5658}{1801}$$

$$< \frac{6013}{1914} < \frac{6368}{2027} < \frac{6723}{2140} < \frac{7078}{2253} < \frac{7433}{2366} < \frac{7788}{2479} < \frac{8143}{2592} < \frac{8498}{2705}$$

$$< \frac{8853}{2818} < \frac{9208}{2931} < \frac{9653}{3044} < \frac{9918}{3157} < \frac{355}{113}\ . \tag{2.6}$$

Except for

$$\frac{2108}{671} = \frac{62 \times 68}{22 \times 61}$$

none of the fractions in (2.6) can be expressed in the form $\frac{a\,b}{c\,d}$ where a, b, c, d are integers between 20 and 100. Since

$$\frac{2108}{671} < \pi,$$

there is no need to consider any fractions and Farey means that are less than $\frac{2108}{671}$. The following is a list of all Farey means, obtainable from $\frac{333}{106}$ and $\frac{355}{113}$, that are $\geq \frac{2108}{671}$ and whose numerator and denominator are $\leq 10{,}000$. Note that this list can be derived by taking Farey means of the fractions listed in (2.6) that are $\geq \frac{2108}{671}$. We have

$$\frac{2108}{671} < \frac{8787}{2797} < \frac{6679}{2126} < \frac{4571}{1455} < \frac{7034}{2239} < \frac{9497}{3023} < \frac{2463}{784} < \frac{7744}{2465}$$

$$< \frac{5281}{1681} < \frac{8099}{2578} < \frac{2818}{897} < \frac{8809}{2804} < \frac{5591}{1907} < \frac{9164}{2917} < \frac{3173}{1010}$$

$$< \frac{9874}{3143} < \frac{6701}{2133} < \frac{3528}{1123} < \frac{7411}{2359} < \frac{3883}{1236} < \frac{8121}{2585} < \frac{4238}{1349}$$

$$< \frac{8831}{2811} < \frac{4593}{1462} < \frac{9541}{3037} < \frac{4948}{1575} < \cdots < \frac{355}{113},$$

where the Farey means between $\frac{4948}{1575}$ and $\frac{355}{113}$ are already given in (2.6). None of the above fractions can be expressed in the required form except for

$$\frac{7744}{2465} = \frac{88 \times 88}{29 \times 85}.$$

Since

$$\left| \frac{7744}{2465} - \pi \right| > \left| \frac{355}{113} - \pi \right|,$$

the search has to be expanded to the interval $[\frac{333}{106}, \frac{22}{7}]$.

3. Since

$$\frac{7744}{2465} < \pi < \frac{355}{113},$$

it suffices to search in the interval $[\frac{355}{113}, \frac{22}{7}]$. We determine the Farey means of the form

$$\frac{(k \times 355) + 22}{(k \times 113) + 7}$$

with numerator and denominator $\leq 10,000$. They are

$$\frac{355}{113} < \frac{9962}{3171} < \frac{9607}{3058} < \frac{9252}{2945} < \frac{8897}{2832} < \frac{8542}{2719} < \frac{8187}{2606} < \frac{7832}{2493}$$

$$< \frac{7477}{2380} < \frac{7122}{2267} < \frac{6767}{2154} < \frac{6412}{2041} < \frac{6057}{1928} < \frac{5702}{1815} < \frac{5347}{1702} < \frac{4992}{1589}$$

$$< \frac{4637}{1476} < \frac{4282}{1363} < \frac{3927}{1250} = \frac{11 \times 355 + 22}{11 \times 113 + 7} < \dots < \frac{22}{7} \qquad (2.7)$$

None of the fractions $< \frac{3927}{1250}$ in (2.7) can be expressed in the required form, except for

$$\frac{3927}{1250} = \frac{51 \times 77}{25 \times 50}.$$

Since

$$\pi < \frac{3927}{1250},$$

there is no need to consider fractions larger than $\frac{3927}{1250}$. The Farey means that are obtainable from $\frac{355}{113}$ and $\frac{22}{7}$, that are $\leq \frac{3927}{1250}$, whose numerator and denominator are $\leq 10,000$ and that are not included in (2.7), are listed below. Note that they can be derived by taking Farey means of the fractions in (2.7) whose numerators are $\leq 5,000$. We have

$$\frac{4992}{1589} < \frac{9629}{3065} < \frac{4637}{1476} < \frac{8919}{2839} < \frac{4282}{1363} < \frac{8209}{2613} < \frac{3927}{1250}.$$

None of the new fractions can be expressed in the required form. Since

$$\left| \frac{3927}{1250} - \pi \right| < \left| \frac{7745}{2465} - \pi \right|,$$

the solution to our problem is a $= 51$, b $= 77$, c $= 25$ and d $= 50$.

Remark.   The computational process can be written easily in the language ALGOL 60, and the above example was completed in 90 seconds on the computer DJS-21 at the Institute of Mathematics, Academia Sinica.

References.

Chen De Quan, Wu Fang and Qi Jing Tai.   "On a solution of the Meshing Gear—Pair Problem."   Math. Recognition and Practice, 3, 1975, pp. 30—40. [Editor's note: The name of the journal is "Shuxue De Shijian Yu Renshi," which has been translated in the past as "Mathematics: Its Cognition and Practice," "Mathematics in Practice and Theory" and "Knowledge, Practice and Mathematics."   The abbreviation currently used in Mathematical Reviews is Math. Practice Theory.]

Hardy, G.H., and E.M. Wright.   An Introduction to the Theory of Numbers. 4th ed. Clarendon Press, Oxford, 1960.

Hua Loo Keng. Introduction to Number Theory. Science Press, Beijing, 1956, Springer Verlag, 1982.

Hua Loo Keng. Introduction to Higher Mathematics. Vol. I. Science Press, Beijing, 1963.

CHAPTER 3

OPTIMUM SEEKING METHODS (single variable)

§ 3.1 Introduction.

An <u>optimum</u> <u>seeking</u> <u>method</u> (optimal search method, You Xuan Fa in Chinese) is a method to find technological production processes that are best in some sense, while using as few experiments as possible. It is a scientific method for arranging experiments. From our contacts with a large number of industries over the long period from 1970 to 1982, we have learned that this topic offers some of the most appropriate techniques for popularization.

The golden section method, the Fibonacci search, the bisection method and the parabola method are discussed in this chapter and they have been popularized widely. Each method takes only one factor of a given production process into consideration and obtains a better production technology for this factor. Then the method can be repeated for other factors. Optimum seeking methods that treat several variables simultaneously will be discussed in the next chapter.

Now we list some of the real problems we have tackled.

1. In chemical engineering: To seek the optimal combination (and their ratios) of three types of liquid crystal from 200 different kinds, so as to satisfy the color and sensitivity requirements.

2. In the petroleum industry: To seek the optimal pressure, temperature and flow values in a refinery tower for the separation of various components of petroleum.

3. In coal mining: To seek the optimal depth of gunpowder holes

and the optimal distance between two adjacent holes on the excavation surface in order to achieve the desired explosion result with minimum gunpowder consumption.

4. In machine building:

a. Metal cutting: To seek the optimal spindle speed and feed rate, and to seek the optimal angles on the cutting tool.

b. Welding: To seek the optimal electrical current to obtain the best quality for the welded surface and to reduce the welding material consumption.

5. In the electronics industry:

a. Manufacturing of semiconductor components: To seek the optimal technological parameters, such as temperature of die detachment (two-thirds of the melting point is not good enough).

b. Adjusting and testing of new equipment: To minimize the number of tests needed to give the required results.

6. In the textile industry: To seek the optimal composition and consumption of different basic dye ingredients so as to improve the quality of dye-stuffs.

7. In metallurgy: To seek the optimal electric current for the iron smelters and to seek the best technological process for smelting high grade steel and for extracting and purifying Vanadium.

8. In power supply: For the boilers, to seek the optimal diameter of nozzles and the optimal angle of injection so as to increase thermal efficiency and to decrease air pollution.

9. In construction materials: In the manufacturing of cement, to seek the optimal rpm of the container in order to decrease the cycle time and to increase the output.

10. In the food industry: To improve the production techniques of making candy and pastry.

§ 3.2 Unimodal functions.

We assume that $f(x)$ is a <u>unimodal function</u> on the interval (a, b), i.e. $f(x)$ is a continuous function and there exists $x_0 \in (a, b)$ such that $f(x)$ is nondecreasing (or nonincreasing) on $(a, x_0]$ and nonincreasing (nondecreasing) on $[x_0, b)$. The point x_0 is called a maximum (or minimum) point of $f(x)$. Without loss of generality we assume throughout this chapter that $f(x)$ is a unimodal function with an interior maximum point. Note that an expression for the function $f(x)$ is not necessarily known, but its value at any point x can be obtained by experimentation. What is the most effective way to determine the location of a maximum point (or a nearby point) for $f(x)$?

The simplest method is the <u>equi-distribution</u> <u>method</u>, i.e. the experiments are arranged at

$$x_k = a + \frac{k}{n+1}(b - a), \qquad 1 \le k \le n,$$

and then the values $y_k = f(x_k)$, $1 \le k \le n$, are obtained. Suppose that y_m is largest among the y_k's. Then $f(x)$ has a maximum in the small interval

$$\left(a + \frac{m-1}{n+1}(b - a), a + \frac{m+1}{n+1}(b - a) \right).$$

The number of trials necessary for the equi-distribution method is too large. More precisely, if we want to obtain a small interval of length less than ϵ that contains a maximum point of $f(x)$, then the number of trials needed is of order of magnitude $\frac{1}{\epsilon}$.

§ 3.3 Methods of trials by shifting to and fro.

We often use the method of trials by shifting to and fro. Take a point x_1 and obtain a value $y_1 = f(x_1)$ by an experiment. Then take another point x_2 ($> x_1$) and obtain $y_2 = f(x_2)$ (see Figure 3.1).

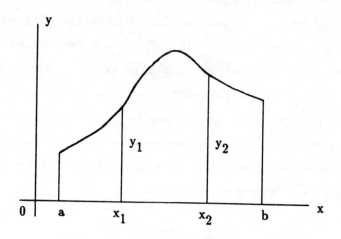

Figure 3.1

If $y_2 > y_1$, then the maximum points of $f(x)$ are obviously not in the interval $(a, x_1]$, so it suffices to find a maximum point of $f(x)$ in the interval (x_1, b). Similarly, if $y_2 < y_1$, then one only needs to consider the interval (a, x_2). In case $y_2 = y_1$ one may proceed with either (x_1, b) or (a, x_2). Without loss of generality we suppose that $y_2 > y_1$. Take a third trial point x_3 in (x_1, b) and obtain $y_3 = f(x_3)$. Then y_2 and y_3 are compared as before and the appropriate part of the interval (x_1, b) is eliminated. This procedure is repeated until a satisfactory approximation of a maximum point of $f(x)$ is found.

§ 3.4 The golden section method.

A natural question is whether there is any rule to select the trial points x_1, x_2, ... so as to find the maximum point of $f(x)$ as quickly as possible. This problem was solved by J. Kiefer in 1953. Since his method is connected with the golden number of elementary geometry, it is called the golden section method. Let

$$\omega = \frac{\sqrt{5} - 1}{2} = 0.6180339887\ldots\,,$$

which is called the golden number. It satisfies the quadratic equation

$$\omega^2 + \omega - 1 = 0.$$

Without loss of generality we may suppose that the interval for experimentation is $(0, 1)$. The golden section method may be described as follows. Arrange the first trial at ω and the second trial at $1 - \omega$, the point opposite ω (see Figure 3.2).

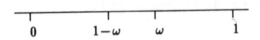

Figure 3.2

Compare the results at these two points. If the result at ω is better, then eliminate the interval $(0, 1 - \omega]$. Otherwise we eliminate the interval $[\omega, 1)$. In either case the length of the remaining interval is equal to $1 - \omega$, i.e. ω multiplied with the length of the original interval. The remaining interval always includes a trial point (ω or $1 - w$) that divides the interval into two parts whose ratio is the same as the ratio of the two parts in which ω or $1 - \omega$ divide the interval

$(0,1)$. More specifically, if the eliminated interval is $[\omega,1)$, then the remaining interval is

$$0 < 1 - \omega = \omega^2 < \omega,$$

which is ω multiplied with the interval

$$0 < \omega < 1.$$

Otherwise, if the eliminated interval is $(0, 1-\omega]$, then the remaining interval is

$$1 - \omega < (1 - \omega) + \omega(1 - \omega) < (1 - \omega) + \omega = 1,$$

which is ω multiplied with the interval

$$0 < 1 - \omega < 1,$$

together with a shift to the right of size $1 - \omega$. Hence the length of the remaining interval after the n-th trial is equal to

$$g_n = \omega^{n-1}.$$

This means that if, using the golden section method, we want to find an interval with length smaller than ϵ that contains a maximum point of the function f, then the smallest number n of trials needed satisfies

$$\omega^{n-1} < \epsilon < \omega^{n-2},$$

i.e. n is of order of magnitude $\ln \frac{1}{\epsilon}$. This is much better than the result for the equi—distribution method.

Let $\mathcal{U}$ denote the set of unimodal functions on $(0,1)$ with interior maximum points. For any $f \in \mathcal{U}$ and any method A of trials by shifting to and fro, let $\Delta_n(A, f)$ denote the length of the remaining interval (containing a maximum point of f) after the n-th trial. Furthermore, let

$$\Delta_n(A) = \max_{f \in \mathcal{U}} \Delta_n(A, f),$$

which is called the Δ-error of the method A after the n-th trial. It is evident that $\Delta_n(A) = g_n$ when A is the golden section method.

Theorem 3.1. For any given method A of trials by shifting to and fro that is different from the golden section method, there exists a positive integer N such that for all $n > N$ the error $\Delta_n(A)$ satisfies

$$\Delta_n(A) > g_n .$$

Remark. Consider the example in section 0.4. If we use 0.618 instead of the golden number ω, then only 16 trials are needed to determine the best temperature. For instance, suppose that $1001°C$ is the best temperature. Then the trials will be arranged at $1618°C$, $1382°C$, $1236°C$, $1146°C$, $1090°C$, $1056°C$, $1034°C$, $1022°C$, $1012°C$, $1010°C$, $1002°C$, $1008°C$, $1006°C$, $1004°C$, $1003°C$ and $1001°C$ successively.

§ 3.5 The proof of Theorem 3.1.

Let F_n $(n = 0, 1, \ldots)$ be the sequence of integers defined recursively by

$$F_0 = 0, \quad F_1 = 1, \quad F_n = F_{n-1} + F_{n-2} \quad (n \geq 2).$$

This is called the <u>Fibonacci sequence</u> and F_n is the n-th Fibonacci number.

Lemma 3.1. We have

$$F_n = \frac{1}{\sqrt{5}} (\omega^{-n} - (-\omega)^n), \qquad n = 0, 1, \ldots.$$

Proof. Since $\omega^{-1} = (\sqrt{5} + 1)/2$, the statement is true for $n = 0$ and $n = 1$. Suppose that $n \geq 2$ and that the lemma holds for all nonnegative integers less than n. Then

$$F_n = F_{n-1} + F_{n-2}$$

$$= \frac{1}{\sqrt{5}} (\omega^{-n+1} - (-\omega)^{n-1} + \omega^{-n+2} - (-\omega)^{n-2})$$

$$= \frac{1}{\sqrt{5}} (\omega^{-n+1}(1 + \omega) - (-\omega)^{n-1}(1 - \omega^{-1}))$$

$$= \frac{1}{\sqrt{5}} (\omega^{-n} - (-\omega)^n),$$

and the lemma follows by induction. □

Let A be a method of trials by shifting to and fro. Write Δ_n instead of $\Delta_n(A)$.

Lemma 3.2. We have

$$\Delta_n = F_{k+1}\Delta_{n+k} + F_k\Delta_{n+k+1}, \quad n = 1, 2, \ldots, \quad k = 0, 1, \ldots. \quad (3.1)$$

Proof. (3.1) holds obviously for $k = 0$. For the case $k = 1$ we need to prove that

$$\Delta_n \leq \Delta_{n+1} + \Delta_{n+2} \qquad n = 1, 2, \ldots . \qquad (3.2)$$

We will prove (3.2) only for $n = 1$, since the proof for $n > 1$ is similar. Suppose that the first two trials are at x_1 and x_2, where $x_1 < x_2$. If $x_1 + x_2 \geq 1 = \Delta_1$, then $\Delta_2 \geq x_2$ and we have $\Delta_3 \geq x_1$ for any choice of the third trial point x_3. Hence

$$\Delta_1 = 1 \leq x_2 + x_1 \leq \Delta_2 + \Delta_3.$$

Otherwise, if $x_1 + x_2 < 1$, then $\Delta_2 \geq 1 - x_1$ and $\Delta_3 \geq 1 - x_3$ for any choice of the third trial point x_3. Hence

$$\Delta_2 + \Delta_3 \geq (1 - x_1) + (1 - x_2) = 2 - (x_1 + x_2) > 1 = \Delta_1,$$

and this proves (3.2) for $n = 1$.

Suppose that $k \geq 1$ and that (3.1) is true for k and all integers $n \geq 1$. We proceed to prove that (3.1) holds also for $k+1$ and all $n \geq 1$. From (3.2) we have

$$\Delta_n \leq F_{k+1}\Delta_{n+k} + F_k \Delta_{n+k+1}$$

$$\leq F_{k+1}(\Delta_{n+k+1} + \Delta_{n+k+2}) + F_k \Delta_{n+k+1}$$

$$= (F_{k+1} + F_k)\Delta_{n+k+1} + F_{k+1}\Delta_{n+k+2}$$

$$= F_{k+2}\Delta_{n+k+1} + F_{k+1}\Delta_{n+k+2}$$

and the lemma follows by induction. □

<u>Lemma</u> 3.3. We have

$$g_n = F_{k+1} g_{n+k} + F_k g_{n+k+1}, \quad n = 1, 2, \ldots, \quad k = 0, 1, \ldots . \quad (3.3)$$

Proof. (3.3) holds clearly for $k = 0$. Since

$$\omega^{n-1}(\omega^2 + \omega - 1) = 0,$$

i.e.

$$g_n = g_{n+1} + g_{n+2},$$

(3.3) is true for $k = 1$, and the lemma is now easily proved by induction. □

The <u>proof</u> of Theorem 3.1. If the theorem is not true, then there is a method A of trials by shifting to and fro that is different from the golden section method and there is an infinite sequence of positive integers

$$n_1 < n_2 < \ldots$$

such that

$$\Delta_{n_i} \leq g_{n_i}, \quad i = 1, 2, \ldots . \quad (3.4)$$

For any given positive integer n there exists n_i such that $n < n_i$. We may suppose without loss of generality that $i = 1$ and will write $n_0 = n$. From Lemma 3.2 we have

$$\Delta_n = \Delta_{n_0} \leq F_{n_1 - n_0 + 1} \Delta_{n_1} + F_{n_1 - n_0} \Delta_{n_1 + 1}$$

Let $s \geq 2$. Since

$$\Delta_{n_i+1} \leq F_{n_{i+1}-n_i} \Delta_{n_{i+1}} + F_{n_{i+1}-n_i-1} \Delta_{n_{i+1}+1}, \quad i = 1, \ldots, s-1$$

we have

$$\Delta_n \leq F_{n_1-n_0+1} \Delta_{n_1} + F_{n_1-n_0} (F_{n_2-n_1} \Delta_{n_2} + F_{n_2-n_1-1} \Delta_{n_2+1})$$

$$\leq \cdots$$

$$\leq F_{n_1-n_0+1} \Delta_{n_1} + F_{n_1-n_0} F_{n_2-n_1} \Delta_{n_2} + \cdots$$

$$+ F_{n_1-n_0} F_{n_s-n_{s-1}} \prod_{i=1}^{s-2} F_{n_{i+1}-n_i-1} \Delta_{n_s}$$

$$+ F_{n_1-n_0} \prod_{i=1}^{s-1} F_{n_{i+1}-n_i-1} \Delta_{n_s+1}$$

Similarly, from Lemma 3.3 we have

$$g_n \leq F_{n_1-n_0+1} g_{n_1} + F_{n_1-n_0} F_{n_2-n_1} g_{n_2} + \cdots$$

$$+ F_{n_1-n_0} F_{n_s-n_{s-1}} \prod_{i=1}^{s-2} F_{n_{i+1}-n_i-1} g_{n_s}$$

$$+ F_{n_1-n_0} \prod_{i=1}^{s-1} F_{n_{i+1}-n_i-1} g_{n_s+1}$$

Since (3.4) and

$$\Delta_{n_s+1} \leq \Delta_{n_s} \leq g_{n_s},$$

it follows from Lemma 3.3 that

$$\Delta_n - g_n \le F_{n_1-n_0} \prod_{i=1}^{s-1} F_{n_{i+1}-n_i-1}(g_{n_s} - g_{n_s+1})$$

$$= F_{n_1-n_0} \prod_{i=1}^{s-1} F_{n_{i+1}-n_i-1} g_{n_s+2}$$

By Lemma 3.1 we have

$$F_n \le \frac{\omega^{-n+1}}{\sqrt{5}}(\omega^{-1} + \omega) = \omega^{-n+1},$$

and consequently

$$\Delta_n - g_n \le \omega^{-n_1+n_0+1} \prod_{i=1}^{s-1} \omega^{-n_{i+1}+n_i+2} \omega^{n_s+1}$$

$$= \omega^{n+2s}.$$

Let $s \to \infty$. Then

$$\Delta_n \le g_n, \quad n = 1, 2, \dots.$$

Since the method A is not the golden section method, there exists a positive integer m such that

$$\Delta_m < g_m.$$

It follows from Lemmas 3.2 and 3.3 that

$$1 = \Delta_1 \le F_m \Delta_m + F_{m-1} \Delta_{m+1}$$

$$< F_m g_m + F_{m-1} g_{m+1} = g_1 = 1,$$

which is a contradiction. This proves the theorem. □

§ 3.6 The Fibonacci search.

Sometimes the parameters in a technological process do not vary continuously. A lathe, say, has only a limited number of turning speeds, so it is difficult to apply the number

$$\omega = \frac{\sqrt{5} - 1}{2} = 0.6180339887\ldots.$$

or 0.618. The theory of continued fractions (see Chapter 2) can play an important role in such cases. The expansion of ω into the simple continued fraction is

$$\omega = [0, 1, 1, \ldots, 1, \ldots]$$

and its convergents are

$$\frac{0}{1}, \frac{1}{1}, \frac{1}{2}, \frac{2}{3}, \ldots, \frac{F_n}{F_{n+1}}, \ldots$$

Suppose there are $F_{n+2} - 1$ points labeled

$$1, 2, \ldots, F_{n+2} - 1 \tag{3.5}$$

that can be used as the trial points. Similar to the golden section method but now with

$$\frac{F_{n+1}}{F_{n+2}}$$

instead of ω, select the first trial point at F_{n+1} and the second trial point at the point opposite the first point, i.e. at $F_{n+2} - F_{n+1} = F_n$. Then compare the results at these two points. If the result at

F_{n+1} is better than that at F_n, then eliminate the points $1, \ldots, F_n$. Otherwise cancel $F_{n+1}, \ldots, F_{n+2} - 1$. We may suppose that F_n gives a better result. Then arrange the third trial at the position that is symmetric to F_n, i.e. at

$$F_{n+1} - F_n = F_{n-1},$$

etcetera, until a maximum point of the unimodal function on (3.5) is found using n trials.

If the number of available trial points is m and

$$F_{n+1} \leq m < F_{n+2} - 1,$$

then we may add $F_{n+2} - m - 1$ slack points

$$m + 1, m + 2, \ldots, F_{n+2} - 1$$

and pretend that the experimental results at these points are all equal to the result at m. Hence this situation is reduced to the case of $F_{n+2} - 1$ points. This method to determine an optimum for a unimodal function on a finite ordered set is called the Fibonacci search.

For any method A of trials by shifting to and fro and any $f \in \mathcal{U}$, let $\phi_n(A, f)$ be the maximum number of trial points for which method A can determine the maximum point of f using n trials. Let

$$\phi_n(A) = \underset{f \in \mathcal{U}}{\text{MIN}} \; \phi_n(f, A).$$

It is called the ϕ-number of the method A after n trials.

We denote by f_n the ϕ-number of the Fibonacci search after n trials. Then

$$f_n = F_{n+2} - 1$$

and we have the following theorem.

<u>Theorem</u> 3.2. Let

$$\Phi_n = \max_A \phi_n(A).$$

Then

$$\Phi_n = f_n .$$

§ 3.7 The proof of Theorem 3.2.

If $n = 1$, then $\Phi_n = 1$. If $n = 2$, then the maximum point of any f can be determined with two experiments and thus $\Phi_2 = 2$. For the general case, suppose that B is a method of trials by shifting to and fro and that $g \in \mathcal{U}$ such that

$$\Phi_n = \phi_n(B) = \phi_n(B, g).$$

Let $f \in \mathcal{U}$ be an arbitrary function whose domain contains Φ_n points. Suppose that the method B prescribes that the first two trials are made at the points x_1 and x_2 among the Φ_n points, where x_1 is located at the $(a+1)$-th point from the left, x_2 at the $(a+b+2)$-th point and that there are c points to the right of x_2 as shown in Figure 3.3.

$$\begin{array}{ccccccccccccc} \circ\circ & \cdots & \circ\circ & \bigcirc & \circ\circ & \cdots & \circ\circ & \bigcirc & \circ\circ & \cdots & \circ\circ \\ & \longleftarrow a \longrightarrow & & x_1 & & \longleftarrow b \longrightarrow & & x_2 & & \longleftarrow c \longrightarrow & \end{array}$$

Figure 3.3

Thus

$$\Phi_n = a + b + 2. \tag{3.6}$$

If the result at x_1 is better than that at x_2, then a maximum point of the function f occurs among the points shown in Figure 3.4.

oo ··· oo O oo ··· oo

⟵ a ⟶ x_1 ⟵ b ⟶

Figure 3.4

With method B we are able to determine the maximum of the function f on the domain given in Figure 3.4 in $n-1$ trials, where the trial at x_1 is included. Thus

$$a + b + 1 \leq \phi_{n-1}(B, f)$$

and, since f is arbitrary,

$$a + b + 1 \leq \phi_{n-1}(B) \leq \Phi_{n-1}. \tag{3.7}$$

Since the number of trials employed by method B on the leftmost a points in Figure 3.4 is at most $n-2$, we have

$$a \leq \phi_{n-2}(B, f)$$

and, since f is arbitrary,

$$a \leq \phi_{n-2}(B) \leq \Phi_{n-2}. \tag{3.8}$$

Similarly, employing functions for which x_2 gives the better result, it follows that

$$b + c + 1 \leq \Phi_{n-1} \quad \text{and} \quad c \leq \Phi_{n-2}. \qquad (3.9)$$

Hence it follows from (3.6), (3.7), (3.8) and (3.9) that

$$\phi_n \leq \phi_{n-1} + \phi_{n-2} + 1.$$

Equality holds if and only if equalities hold in (3.7), (3.8) and (3.9), which is equivalent to the conditions

$$a = c = \Phi_{n-2} \quad \text{(symmetry)}$$

and

$$a + b + 1 = b + c + 1 = \Phi_{n-1}.$$

Write the equality

$$\Phi_n = \Phi_{n-1} + \Phi_{n-2} + 1$$

in the form of the recurrence relation

$$\Phi_n + 1 = (\Phi_{n-1} + 1) + (\Phi_{n-2} + 1).$$

Define

$$F_0 = 0, \quad F_1 = F_2 = 1 \quad \text{and} \quad F_{n+2} = \Phi_n + 1 \ (n \geq 1).$$

Then

$$F_0 = 0, \quad F_1 = 1, \quad F_n = F_{n-1} + F_{n-2} \ (n \geq 2),$$

i.e. F_n $(n = 0, 1, \dots)$ is the Fibonacci sequence, and the theorem is proved. $\square$

The Fibonacci search can also be used in case the variable varies continuously. In particular, the use of the Fibonacci search is convenient if the number of trials is a preassigned number. More precisely, we may divide the interval $(0, 1)$ into F_{n+2} equal intervals and use $\dfrac{F_{n+1}}{F_{n+2}}$ instead of 0.618. The first trial is taken at $\dfrac{F_{n+1}}{F_{n+2}}$ and the second trial at

$$1 - \frac{F_{n+1}}{F_{n+2}} = \frac{F_n}{F_{n+2}},$$

which is symmetric to $\dfrac{F_{n+1}}{F_{n+2}}$ and so on. After n trials we have a remaining interval with length $\dfrac{2}{F_{n+2}}$ which contains a maximum point of f. Hence the Δ-error of the Fibonacci search after n trials is

$$\Delta_n = \frac{2}{F_{n+2}}.$$

Remark. Consider the example given in section 0.4. Since $F_{17} = 1597$, 15 trials are enough to determine the best temperature between $1000°C$ and $2000°C$, and this number is one trial less than is needed for the golden section method (see section 3.4).

§ 3.8 The bisection method.

We have introduced two methods for finding an optimum of a unimodal function on an interval. But in practice there are many problems where the outcome of an experiment is one of two possible states: yes and no. For instance:

— How can we find as quickly as possible a break in a circuit, a sewer, a petroleum pipeline, a machine or an instrument?

— How can we determine the static equilibrium of a grinding wheel or a rotor?

— How can the consumption of raw materials be reduced without affecting the quality of the products?

— How can the production costs be reduced without affecting the products?

To tackle such problems we suggest using the <u>bisection</u> <u>method</u> instead of the golden section method.

For example, suppose there is a break in a 10 kilometer long electric wire with current flowing at endpoint A but not at the other endpoint B. If we want to find the trouble spot as quickly as possible, then we examine first the midpoint C of the wire AB. Suppose that there is current at C. We conclude that the break occured in CB and we then examine the midpoint D of CB. Assuming that there is no current at D we conclude that the trouble occured in section CD and we then examine the midpoint E of CD, and so on, until the place of trouble is spotted (see Figure 3.5).

Figure 3.5

Now we give another example. Suppose a certain material is needed in a product (its quantity is denoted by x) and suppose there is a function f(x) such that

$f(x) \geq 0$ if the product is up to standard, and

$f(x) < 0$ otherwise.

If we know that $f(x_0) < 0$ and $f(x_1) \geq 0$, where $x_0 < x_1$, then we may test next at $\dfrac{(x_0 + x_1)}{2}$, and so on.

The bisection method is quite simple and is being popularized on a large scale. Suppose the length of the original trial interval is L. Then the length of the remaining interval is $\frac{L}{2}$ after one trial has been done and consequently, the length of the remaining interval is $\dfrac{L}{2^n}$ if n trials have been done. It is clear that the order of magnitude of the number of trials for the bisection method is $\ln \frac{1}{\delta}$, where δ is a preassigned precision, that is, the ratio of the length of a satisfactory interval containing the break point over the length of the original interval.

§ 3.9 The parabola method.

Both the golden section method and the Fibonacci search compare two experimental results at every iteration. However, the actual values of the experiments, the values of the objective function, are not used. We will now suggest an effective method for finding a maximum point of a function that does use the values of the experimental outcomes. For example, suppose that y_1, y_2, y_3 are the values of the function $f(x)$ obtained by trials located at x_1, x_2, x_3 respectively, where $x_1 < x_2 < x_3$. Then obtain the quadratic function

$$y = y_1 \frac{(x - x_2)(x - x_3)}{(x_1 - x_2)(x_1 - x_3)} + y_2 \frac{(x - x_3)(x - x_1)}{(x_2 - x_3)(x_2 - x_1)} +$$

$$+ y_3 \frac{(x - x_1)(x - x_2)}{(x_3 - x_1)(x_3 - x_2)}$$

by Lagrange interpolation, where $y = y_i$ if $x = x_i$. The graph of this function is a parabola and its maximum point is given by the equation

$$y_1 \frac{2x - x_2 - x_3}{(x_1 - x_2)(x_1 - x_3)} + y_2 \frac{2x - x_3 - x_1}{(x_2 - x_3)(x_2 - x_1)}$$

$$+ y_3 \frac{(x - x_1)(x - x_2)}{(x_3 - x_1)(x_3 - x_2)} = 0$$

i.e.

$$x_4 = \frac{y_1(x_2{}^2 - x_3{}^2) + y_2(x_3{}^2 - x_1{}^2) + y_3(x_1{}^2 - x_2{}^2)}{2\left(y_1(x_2 - x_3) + y_2(x_3 - x_1) + y_3(x_1 - x_2)\right)}$$

Take the next trial point at x_4. Suppose that the result at x_4 is y_4 and that the maximum of y_1, y_2, y_3 and y_4 is reached at $x_i{}'$. Besides $x_i{}'$, take the two points among x_1, x_2, x_3 and x_4 that are nearest to $x_i{}'$. Denote these three points by $x_1{}', x_2{}'$ and $x_3{}'$, where $x_1{}' < x_2{}' < x_3{}'$. Denoting the function values at $x_1{}', x_2{}', x_3{}'$ by $y_1{}', y_2{}', y_3{}'$ respectively, we can now obtain a new parabola and determine its maximum point. Continue in this manner until a maximum point (or a nearby point) of the function is found. This method is called the <u>parabola method</u> (or quadratic interpolation) and its rate of convergence is often better than that of the golden section method (Hua [1981]).

In practice one may use the golden section method first and then use its last three trial points x_1, x_2, x_3 and the corresponding values y_1, y_2, y_3 to find x_4 and the corresponding y_4 by the parabola method. Compare y_4 and the maximum of y_1, y_2 and y_3 and then decide whether to make a trial at x_4.

References.

Hua Loo Keng. Popular Lectures on Optimum Seeking Methods. Science Press, Beijing, 1971.

Hua Loo Keng. Popular Lectures on Optimum Seeking Methods (with Supplements). Guo-Fang Industry Press, Beijing, 1973.

Hua Loo Keng. Some Popular Lectures on Optimum Seeking Methods. Liao Ning People Press, Liao Ning, 1973.

Hua Loo Keng. Theory of Optimization. Science Press, Beijing, 1981.

Hua Loo Keng, Chen De Quan, Ji Lei etc (edited). A Collection of Literatures on Overall Planning Methods. Chinese Univ. of Science and Tech. Press, 1965.

Kiefer J. "Sequential Minimax Search for a Maximum" Proc. Amer. Math. Soc, 4, 1953, pp. 502-506.

Wilde, D.J. Optimum Seeking Methods. Prentice Hall, N.J., 1964.

Wu Fang. "Extremum of a Special Function" Scientia Sinica, 1, 1974, pp. 1-14.

Editor's note: Some additional references for this area are the following.

Avriel, M. Nonlinear Programming: Analysis and Methods. Prentice-Hall, 1976.

Bazaraa M.S., and C.M. Shetty. Nonlinear Programming. John Wiley & Sons, 1979.

Beightle C.S., D.T. Phillips and D.J. Wilde. Foundations of Optimization. 2nd ed., Prentice-Hall, 1979.

Kowalik J., and M.R. Osborne. "Methods for Unconstrained Optimization Problems" in Modern Analytic and Computational Methods, ed. R. Bellman. Am. Elsevier Publ. Comp., N.Y. 1968.

Press, W.H., B.P. Flannery, S.A. Teukolsky and W.T. Vetterling. Numerical Recipes. Cambridge University Press, 1986.

Scales, L.E. Introduction to Non-Linear Optimization. Springer, 1985.

CHAPTER 4

OPTIMUM SEEKING METHODS (several variables)

§ 4.1 Introduction.

For simplicity, our discussion is confined to the case of two variables only. Most of the optimum seeking methods for two variables can be easily generalized to the case of more than two variables, but the number of required trials grows very fast when the number of variables is increased. Thus the use of optimum seeking methods for several variables should be avoided as much as possible, even though there are always many factors that influence an industrial production process. It is better to grasp the principal factors and to use an optimum seeking method on one or two variables so as to obtain a better production technology for these factors, and then use the same method to improve the production technologies with respect to the next one or two factors, and so on, until a satisfactory result is obtained. It seems that this is more reliable and reasonable than the direct use of an optimum seeking method for several variables. We refer back to section 3.1 for some of the real problems we have tackled.

§ 4.2 Unimodal functions (several variables).

We suppose that $f(x,y)$ is a unimodal function defined on the region

$$a < x < b, \ c < y < d. \tag{4.1}$$

That is to say, $f(x,y)$ is continuous, $f(x,y)$ has only one maximum point (or minimum point) in the domain (4.1), and the region enclosed by any contour $f(x,y) = c$ is convex. A contour map of $z = f(x,y)$ is shown in Figure 4.1.

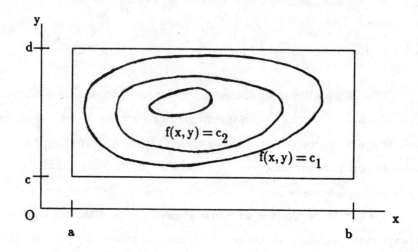

Figure 4.1

The simplest optimum seeking method for two variables is the equi-distribution method. Divide the ranges of x and y in (4.1) into $l+1$ and $m+1$ equal parts. Compare the experimental results at the points

$$(a + \frac{(b-a)i}{l+1}, \ c + \frac{(d-c)j}{m+1}), \quad (1 \le i \le l, 1 \le j \le m)$$

and regard the point with the best result as the approximate location of the maximum point of z.

However, the number of trials needed for this method is too large for a factory. More precisely, if we want to obtain a square with side $\le \epsilon$ that contains the maximum point of f, then the number of trials needed is of order of magnitude $\frac{1}{\epsilon^2}$.

We will give generalizations of the golden section method and the Fibonacci search to the case of two factors in the following sections.

§ 4.3. The bisection method.

1. On the vertical perpendicular bisector of (4.1),

$$x = \frac{a+b}{2}, \quad c < y < d, \tag{4.2}$$

determine with the golden section method the maximum point (or a nearby point) P of the function z, and on the horizontal bisector of (4.1),

$$a < x < b, \quad y = \frac{c+d}{2}, \tag{4.3}$$

determine with the golden section method the maximum point Q of z. Without loss of generality, we may suppose that the locations of P and Q are as shown in Figure 4.2.

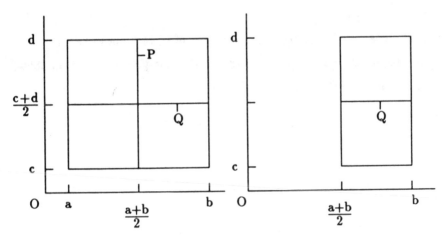

Figure 4.2   Figure 4.3

2. Compare the experimental results at P and Q. If Q gives the better result, then eliminate the region to the left of (4.2) while keeping the other part, as shown in Figure 4.3. Otherwise eliminate the region below (4.3) and keep the part above (4.3), as shown in Figure 4.4.

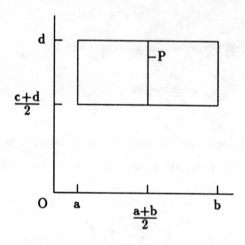

Figure 4.4

3. Without loss of generality, suppose the remaining region is

$$\frac{a+b}{2} < x < b, \quad c < y < d. \tag{4.4}$$

Repeat steps 1 and 2 by using (4.4) instead of (4.1). Since (4.4) includes a trial point Q, it suffices to determine with the golden section method the maximum point R of z on the segment

$$x = \frac{1}{2}(\frac{a+b}{2} + b) = \frac{a+3b}{4}, \quad c < y < d.$$

Then we may again compare the experimental results at Q and R, and so on, until a satisfactory point is found.

We often suppose that the curve $z = f(x, y)$ has continuous tangents. The bisection method is based on the fact that the points for which $f(x, y) > c$ are always located on one side of any tangent line to the curve $f(x, y) = c$, since the region enclosed by the curve $f(x, y) = c$ is convex. It follows that if S is the maximum point of z on a line l, then l is the tangent line to the curve $f(x, y) = c$ at S.

Let $N(\epsilon)$ denote the number of trials used by the bisection method in order to obtain a square region with area $\leq \epsilon^2$ that contains the maximum point of z.

If k trials have been done with the golden section method on the segment (4.3), we obtain an interval containing the maximum point of z on (4.3) with length

$$(b-a)\,\omega^{k-1}, \quad \omega = \frac{\sqrt{5}-1}{2}.$$

Now it is required that

$$(b-a)\,\omega^{k-2} > \epsilon \geq (b-a)\,\omega^{k-1},$$

i.e.

$$(\tfrac{1}{\omega})^{k-1} \geq \tfrac{b-a}{\epsilon} > (\tfrac{1}{\omega})^{k-2}.$$

Hence k satisfies

$$k-1 \geq \frac{\ln\frac{b-a}{\epsilon}}{\ln\frac{1}{\omega}} > k-2,$$

and thus k is approximately equal to

$$\frac{\ln\frac{1}{\epsilon}}{\ln\frac{1}{\omega}}.$$

After k trials have been completed the area of the remaining experimentation region equals one half times the area of the original region. Thus, after n linesegment searches are completed, the area of the remaining region is

$$\frac{(b-a)(d-c)}{2^n} .$$

Let n be the integer satisfying

$$\frac{(b-a)(d-c)}{2^n} \le \epsilon^2 < \frac{(b-a)(d-c)}{2^{n-1}}$$

Then n is approximately equal to

$$\frac{2\left(\ln\frac{1}{\epsilon}\right)}{\ln 2}$$

and therefore $N(\epsilon)$ equals approximately

$$\frac{2\left(\ln\frac{1}{\epsilon}\right)^2}{\ln 2 \ln\frac{1}{\omega}}$$

which is of order of magnitude $\left(\ln\frac{1}{\epsilon}\right)^2$. This is much better than $\frac{1}{\epsilon^2}$, the result for the equi-distribution method.

Remark. If the results at the points P and Q are the same or difficult to distinguish, then P and Q are located on the same contour, as shown in Figure 4.5, and therefore both the left half and the lower half of (4.1), i.e. the shaded part in Figure 4.5, may be eliminated. Consequently, the area of the remaining region

$$\frac{a+b}{2} < x < b, \quad \frac{c+d}{2} < y < d$$

is only one fourth times the area of the original region.

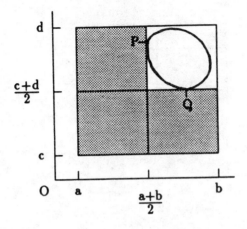

Figure 4.5

§ 4.4 The successive approximation method.

1. Using the golden section method, determine the maximum point (or a nearby point) P_1 of z on the linesegment (4.2), and then determine the maximum point P_2 of z on the horizontal linesegment through P_1. If P_2 is located to the right of P_1, then eliminate the region to the left of (4.2), i.e. the shaded part in Figure 4.6, otherwise eliminate the region to the right of (4.2).

2. Using the golden section method, determine the maximum point P_3 of z on the perpendicular linesegment through P_2. Eliminate the region above the horizontal line through P_2 if P_3 is located below P_2 (see Figure 4.7). Otherwise eliminate the lower part. Continue this process until a satisfactory point is obtained.

The rate of convergence for this method may be derived as follows. Let

$$f(P_i) = z_i, \quad i = 1, 2, \ldots .$$

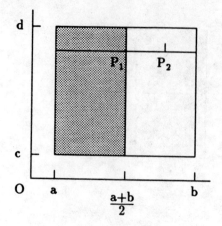

Figure 4.6

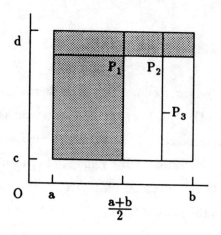

Figure 4.7

Then

$$z_1 < z_2 < \ldots \tag{4.5}$$

is a monotone increasing sequence. Since z_i does not exceed the maximum of the function z, the sequence (4.5) is convergent. Suppose that

$$\lim_{n\to\infty} z_n = z_0.$$

If z_0 is not the maximum of z, then the normal line to the contour l_n at P_n passes through the interior of l_0 for sufficiently large n, where l_0 is the contour with height z_0. That is to say, z_{n+1} is larger than z_0, which is a contradiction. Hence z_0 is the maximum of the function z.

For this method, the ratio of the area of the remaining region and the area of the original region may be either $\geq \frac{1}{2}$ or $< \frac{1}{2}$. So this method is unlike the bisection method, where the area of the remaining region is always equal to one half times the area of the original region.

One may prove (Hua [1981]) that the number of trials needed to determine a rectangle containing the maximum point of z and with area $\leq \epsilon^2$, is also of order of magnitude $(\ln \frac{1}{\epsilon})^2$.

§ 4.5 The parallel line method.

If an experiment is much easier to do with respect to some factor x than with respect to another factor y, then it is convenient to use the so called <u>parallel line method</u>.

Using the golden section method, determine the maximum points P and Q of z on the linesegments

$$a < x < b, \quad y = c + 0.618\,(d - c) \tag{4.6}$$

and

$$a < x < b, \quad y = c + 0.382\,(d - c) \tag{4.7}$$

respectively. Compare the experimental results at P and Q and then eliminate the region below the linesegment (4.7), i.e.

$$a < x < b, \quad c < y < c + 0.382\,(d - c) \tag{4.8}$$

if the result at P is better, otherwise eliminate the region above (4.6), i.e. (see Figure 4.8)

$$a < x < b, \quad c + 0.618\,(d - c) < y < d.$$

Then use the same process on the remaining region, and so on, until a satisfactory point is reached.

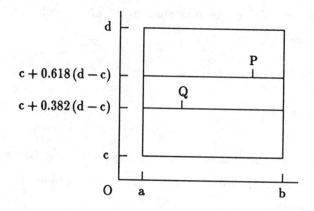

Figure 4.8

It was proved in section 4.3 that in order to obtain for a given linesegment a subinterval of length $\leq \epsilon$ containing the maximum point of z when using the golden section method, the number of necessary trials is of order of magnitude

$$\frac{\ln \frac{1}{\epsilon}}{\ln \frac{1}{\omega}}, \quad \omega = \frac{\sqrt{5} - 1}{2}.$$

The area of the remaining region is always equal to ω times the area of the original region. Hence, after n searches have been completed on n parallel segments, the area of the remaining region is ω^{n-1} times the area of (4.1), i.e. it is equal to

$$(b - a)(d - c)\omega^{n-1}.$$

If this is desired to be close to ϵ^2, then n is approximately

$$\frac{2\ln\frac{1}{\epsilon}}{\ln\frac{1}{\omega}},$$

and therefore the total number of necessary trials is of order of magnitude $(\ln\frac{1}{\epsilon})^2$.

§ 4.6. The discrete case with two factors.

We will illustrate the method by an example. Suppose the function $z = f(x, y)$ is defined on the 20×12 lattice shown in Figure 4.9.

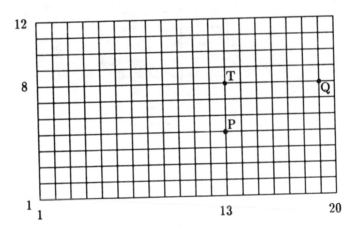

Figure 4.9

1. Using the Fibonacci search, determine in 5 trials the maximum point P of z among the lattice points

$$(13, j) \qquad j = 1, 2, \ldots, 12,$$

and then the maximum point Q among the lattice points

$$(i, 8) \qquad i = 1, 2, \ldots, 20.$$

Since the experimental result at T = (13, 8) is already known, 5 trials are sufficient to determine the maximum Q among the latter group of points. Compare the experimental results at P and Q. If the result at Q is better, then eliminate the region to the left of the vertical segment passing through P, otherwise eliminate the region above the horizontal segment passing through Q. Without loss of generality suppose that Q gives a better result, and so the remaining region is as shown in Figure 4.10.

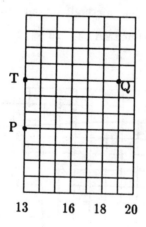

Figure 4.10

2. In Figure 4.10, the Fibonacci search may be used either on the line x = 13 + 5 = 18 or on the line x = 21 − 5 = 16. Since the line

$x = 18$ is closer to Q, we use the Fibonacci search to determine the maximum point R of z among the points

$$(18,j), \quad j = 1, 2, \ldots, 12,$$

and then compare the experimental results at Q and R, and so on. Continuing this process, the maximum point of z is obtained after 30 trials.

In general, we may treat the problem on a $(F_n - 1) \times (F_m - 1)$ lattice. If the number of lattice points N in the horizontal direction is not of the form $F_n - 1$, for example $F_{n-1} - 1 < N < F_n - 1$, then we may add $(F_n - N - 1) \times (F_m - 1)$ lattice points on the right hand side of the region, and define the value of z at these points by

$$f(k,j) = f(N,j), \quad k = N+1, \ldots, F_n - 1, \quad j = 1, 2, \ldots, F_m - 1.$$

In case the number of lattice points in the vertical direction is not of the form $F_m - 1$, we may proceed in a similar manner.

Remark. For demonstrations of the above mentioned methods, it is suggested to prepare a piece of paper with scales like that for the golden section method. A part of the paper is torn away after two experimental results have been compared. Since the maximum point of z is always located on the remaining paper, a satisfactory point is found if the area of the remaining paper is sufficiently small.

§ 4.7 The equilateral triangle method.

Start with any equilateral triangle $\triangle$ ABC in the trial region and arrange the first three trials at the vertices A, B and C. If, say, the result at C is the best one, then prolong the edges AC and BC to AD and BE so that △ CDE and △ ABC are two congruent triangles. Arrange trials at D and at E. If, say, the result at D is the best one among those at C, D and E, then consider the equilateral triangle △ DFG. Continue this process if one of the experimental results at F and G is better than that at D. Otherwise, if D gives a better result than F and G, then consider the midpoints F' and G' of DF and DG, or C' and E' of CD and DE. Continue the above process with triangle △ DF'G' or △ DC'E', until a point O is determined such that no point with a better experimental result than that at O can be found (or until a point is found whose experimental result is difficult to distinguish from those of its neighbours). Then O is the maximum point of z (see Figure 4.11).

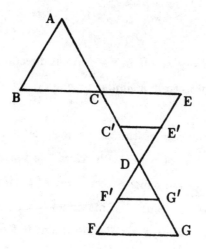

Figure 4.11

The equilateral triangle method is based on the fact that if two directional derivatives at a point P are both equal to zero, then P is usually the maximum point of z.

The restriction to equilateral triangles in this method is clearly unnecessary and one may instead use any given triangle, or more general, any polygon. The simplest case is to use a rectangle instead of a triangle and the method is then called the rectangle method, but since two directions are sufficient for a planar optimization problem it is better to use a triangle rather than a rectangle.

A possible choice for the vertices of the initial equilateral triangle is as follows. Suppose A has coordinates (a_1, a_2). Then the coordinates of the other vertices B and C of the equilateral triangle $\triangle$ ABC with edgelength h may be taken as

$$(a_1 + \frac{\sqrt{3}+1}{2\sqrt{2}}h, \; a_2 + \frac{\sqrt{3}-1}{2\sqrt{2}}h) \quad \text{and} \quad (a_1 + \frac{\sqrt{3}-1}{2\sqrt{2}}h, \; a_2 + \frac{\sqrt{3}-1}{2\sqrt{2}}h)$$

The equilateral triangle method is easily generalized to the case of n variables. Start with a n-dimensional regular simplex with edgelength h, arrange the first $n+1$ trials at the $n+1$ vertices of the simplex and keep the vertex with the best experimental result. Then consider a new regular simplex with edgelength h for which the preserved vertex is one the vertices. Continue the above process until the best result for the new simplex is achieved at the preserved vertex of the old simplex. Then shorten the edgelength of the simplex to $\frac{h}{2}$ and repeat the same process, and so on, until a satisfactory point is found. This method is sometimes called the Simplex Method.

A possible choice for the vertices of the initial regular simplex is as follows. Suppose A has coordinates $(a_1, a_2, \dots, a_n)$. Then the coordinates of the other n vertices of the regular simplex with one vertex at A and edgelength h may be taken as

$$(a_1 + ah, \dots, a_{i-1} + ah, \; a_i + (a + \frac{1}{\sqrt{2}})h, \; a_{i+1} + ah, \dots, a_n + ah),$$

$1 \le i \le n$, where

$$a = \frac{\sqrt{n+1} - 1}{n\sqrt{2}}.$$

§ 4.8 The gradient method.

Let $(x_1, x_2, \ldots, x_s)$, $(x_1 + \delta_1, x_2, \ldots, x_s)$, $\ldots$, $(x_1, x_2, \ldots, x_s + \delta_s)$ be $s+1$ given points in the trial region. Let $z^0, z^1, \ldots, z^s$ be the corresponding experimental results. Determine, using an optimum seeking method for one variable, the maximum point on the linesegment whose points are of the form

$$x_1 + \frac{z^1 - z^0}{\delta_1} t, \ldots, x_s + \frac{z^s - z^0}{\delta_s} t,$$

where the range of t is determined by

1. $t \geq 0$ and

2. $a_i < x_i + \frac{z^i - z^0}{\delta_i} t < b_i, \quad 1 \leq i \leq s.$

(The trial region is $\{u : a_i < u_i < b_i, 1 \leq i \leq s\}$).

If z takes its maximum at $t = t_0$, then repeat the above step with $(y_1, \ldots, y_s)$ instead of $(x_1, \ldots, x_s)$, where

$$y_i = x_i + \frac{z^i - z^0}{\delta_i} t_0, \quad 1 \leq i \leq s.$$

The iteration process stops when the distance between $(x_1, \ldots, x_s)$ and $(y_1, \ldots, y_s)$ is less than a pre-assigned positive number.

If z is assumed to be smooth enough, then one may prove (Hua [1981]) that the number of trials needed to determine a region that contains the maximum point of z and with volume $\leq \epsilon$, is of order of magnitude $(\ln \frac{1}{\varepsilon})^2$.

The gradient method and the successive approximation method have no essential difference, although they are distinct in form. In the latter method, we find first the maximum point P of z on a line l. The line l is the tangent line to the contour $z = c$ at P and thus these two methods are essentially the same.

§ 4.9. The paraboloid method.

The parabola method for a single variable can be sketched as follows. Use the three experimental results y_1, y_2 and y_3 at the points x_1, x_2 and x_3 to determine the parabola that passes through (x_1, y_1), (x_2, y_2) and (x_3, y_3), and then use the maximum point of the parabola as the new trial point and so on. This idea may be generalized to the case of several variables and thus we have the paraboloid method. Given $\frac{(s+1)(s+2)}{2}$ points $P_i (x_1^i, \dots, x_s^i)$, $1 \leq i \leq \frac{(s+1)(s+2)}{2}$, with corresponding experimental results z_i, determine the paraboloid

$$z = F(x_1, \dots, x_s) = \sum_{1 \leq i \leq j \leq s} a_{ij} x_i x_j + \sum_{k=1}^{s} a_k x_k + a \qquad (4.9)$$

passing through $(x_1^i, \dots, x_s^i, z_i)$, $1 \leq i \leq \frac{(s+1)(s+2)}{2}$. Next, find the maximum point $(x_1^*, \dots, x_s^*, \bar{z})$ of (4.9). And finally, determine the new paraboloid passing through $(x_1^i, \dots, x_s^i, z_i)$, $2 \leq i \leq \frac{(s+1)(s+2)}{2}$, and $P^* (x_1^*, \dots, x_s^*, z^*)$, where z^* is the experimental result at $(x_1^*, \dots, x_s^*)$. The iteration process continues until a satisfactory point is found.

Under certain conditions one can obtain an estimate of the rate of convergence for this method (Hua [1981]). More precisely, the number of trials needed to determine a region that contains the maximum point of the function and whose volume is $\leq \epsilon$, is of order of magnitude $s^2 \ln (\ln \frac{1}{\epsilon})$.

Suppose that an expression for z is given, i.e. $z = \phi(x_1, \ldots, x_s)$. Then we can obtain an iterative algorithm to determine the maximum point of z that it is easily converted into a computer program.

Let

$$\vec{x}^0, \; \vec{x}^0 + \eta e_i \; (1 \leq i \leq s), \; \vec{x}^0 + \eta(\vec{e}_i + \vec{e}_j) \; (1 \leq i \leq j \leq s) \qquad (4.10)$$

be $\dfrac{(s+1)(s+2)}{2}$ given points, where $\vec{x}^0 = (x_1^0, \ldots, x_s^0)$ and $\vec{e}_i$ is the i-th unit vector. Determine a quadratic function

$$\Psi(\vec{x}) = \alpha + 2\vec{b}\,\vec{x}^T - \vec{x}\,C\vec{x}^T$$

such that

$$\Psi(\vec{x}) = \phi(\vec{x}) \qquad (4.11)$$

holds for all points of (4.10), where $\vec{b} = (b_1, \ldots, b_s)$ and $C = (c_{ij})$ ($c_{ij} = c_{ji}$) is a $s \times s$ positive definite matrix. Substituting (4.10) into (4.11), we obtain a system of $\dfrac{(s+1)(s+2)}{2}$ equations, and therefore the $\dfrac{(s+1)(s+2)}{2}$ unknowns α, $\vec{b}$ and C are uniquely determined. Let

$$\Delta_i f(\vec{x}) = \frac{1}{\eta} (f(\vec{x} + \eta \vec{e}_i) - f(\vec{x})).$$

Then

$$\Delta_i \phi(\vec{x}^0) = \Delta_i \Psi(\vec{x}^0)$$

$$= \frac{1}{\eta} \left(2\vec{b}\,(\vec{x}^0 + \eta \vec{e}_i)^T - 2\vec{b}\,\vec{x}^{0T} - (\vec{x}^0 + \eta \vec{e}_i)C(\vec{x}^0 + \eta \vec{e}_i)^T - \vec{x}^0 C\vec{x}^{0T} \right)$$

$$= 2\vec{b}\,\vec{e_i}^{\mathrm{T}} - 2\vec{x}^0 C\vec{e_i}^{\mathrm{T}} - \eta\vec{e_i}\,C\vec{e_i}^{\mathrm{T}}.$$

and

$$\Delta_{\underset{i}{}}\Delta_{\underset{j}{}}\phi(\vec{x}^0) = -2\vec{e_i}\,C\vec{e_j}^{\mathrm{T}}.$$

Thus we have

$$C = -\tfrac{1}{2}(\Delta_{\underset{i}{}}\Delta_{\underset{j}{}}\phi(\vec{x}^0)), \quad 1 \le i, j \le s$$

and

$$(\Delta_{\underset{1}{}}\phi(\vec{x}^0), \dots, \Delta_{\underset{s}{}}\phi(\vec{x}^0)) = 2\vec{b} - 2\vec{x}^0 C + \tfrac{\eta}{2}(\Delta_{\underset{1}{}}^2\phi(\vec{x}^0), \dots, \Delta_{\underset{s}{}}^2\phi(\vec{x}^0)).$$

$$(4.12)$$

Since C is positive definite and

$$\Psi(\vec{x}) = \alpha + \vec{b}\,C^{-1}\vec{b}^{\mathrm{T}} - (\vec{x} - \vec{b}\,C^{-1})C(\vec{x} - \vec{b}\,C^{-1})^{\mathrm{T}},$$

$\Psi(\vec{x})$ reaches its maximum at

$$\vec{x} = \vec{x}^1 = \vec{b}\,C^{-1}.$$

By (4.12), we have

$$\vec{x}^1 = \vec{x}^0 - V(\vec{x}^0)M(\vec{x}^0)^{-1},$$

where

$$V(\vec{x}^0) = (\Delta_{\underset{1}{}}\phi(\vec{x}^0) - \tfrac{1}{2}\eta\Delta_{\underset{1}{}}^2\phi(\vec{x}^0), \dots, \Delta_{\underset{s}{}}\phi(\vec{x}^0) - \tfrac{1}{2}\eta\Delta_{\underset{s}{}}^2\phi(\vec{x}^0))$$

and

$$M(\vec{x}^0) = (\Delta_{\underset{i}{}}\Delta_{\underset{j}{}}\phi(\vec{x}^0)), \quad 1 \le i, j \le s.$$

Hence the iteration formula

$$\vec{x}^{r+1} = \vec{x}^r - V(\vec{x}^r)M(\vec{x}^r)^{-1}, \quad r = 0, 1, \dots$$

is suggested to approach the maximum point of $\phi(\vec{x})$, where

$$V(\vec{x}^r) = (\Delta_1 \phi(\vec{x}^r) - \tfrac{1}{2}\eta \Delta_1^2 \phi(\vec{x}^r), \dots, \Delta_s \phi(\vec{x}^r) - \tfrac{1}{2}\eta \Delta_s^2 \phi(\vec{x}^r))$$

and

$$M(\vec{x}^r) = (\Delta_i \Delta_j \phi(\vec{x}^r)), \quad 1 \leq i, j \leq s.$$

Suppose that $\phi(\vec{x})$ reaches its maximum at $\vec{x} = \vec{x}^*$. Then

$$\frac{\partial \phi(\vec{x})}{\partial x_i}\bigg|_{\vec{x} = \vec{x}^*} = 0, \quad 1 \leq i \leq s.$$

Let $\|\vec{x}\|$ denote the sum of the absolute values of the components of $\vec{x}$. Then one can prove (Hua [1981]) that

$$\|\vec{x}^{r+1} - \vec{x}^*\| = O(\|\vec{x}^r - \vec{x}^*\|^2),$$

if $\phi(\vec{x})$ satisfies certain conditions.

Remarks. 1. The rate of convergence for the DFP method (Davidon, Fletcher and Powell, 1962) was originally claimed to be

$$\|\vec{x}^{r+1} - \vec{x}^*\| = o(\|\vec{x}^r - \vec{x}^*\|).$$

Hua Loo Keng proved in the mid 60's that the rate of convergence for this technique can be

$$\|\vec{x}^{r+s} - \vec{x}^*\| = O(\|\vec{x}^r - \vec{x}^*\|^2).$$

When he visited Western Europe in 1979, he was told that this result was also proved by W. Burmeister in 1973. But the number of necessary experiments for the paraboloid method is at most one half of that for the

DFP technique.

2. For the equilateral triangle method, the gradient method and the paraboloid method, a point in the trial region has to be determined that will serve as the initial point for the iteration process. It is important to find a good initial trial point. If it is chosen very close to the maximum point of the function, then a few trials are enough to find the maximum point. We suggest the following way for obtaining an initial trial point. Suppose that the trial domain is the unit cube G_s

$$0 \le x_i \le 1, \quad 1 \le i \le s.$$

Then take a set of points $\vec{x}^j$, $1 \le j \le n$, that is uniformly distributed on G_s (Hua Loo Keng and Wang Yuan [1978]) and arrange a set of n trials at these points. Then compare the experimental results $f(\vec{x}^j)$, $1 \le j \le n$, and choose the point with the highest value as the starting point for the iteration. Moreover, if the function is not unimodal, then these experimental results will indicate the distribution of the function values.

§ 4.10 Convex bodies.

A convex body D in a s-dimensional space is defined as follows. If $\vec{x} \in D$ and $\vec{y} \in D$, then the points on the linesegment connecting $\vec{x}$ and $\vec{y}$ must also belong to D, i.e.

$$\vec{x} + t(\vec{y} - \vec{x}) \in D, \quad 0 \le t \le 1.$$

Theorem 4.1. Any hyperplane passing through the center of gravity of D divides D into two parts such that the ratio M of their volumes satisfies

$$\frac{(\frac{s}{s+1})^s}{1-(\frac{s}{s+1})^s} \leq M \leq \frac{1 - (\frac{s}{s+1})^s}{(\frac{s}{s+1})^s}.$$

We will prove the theorem only for the case $s = 2$, since the argument of the proof can be extended to the case $s > 2$.

The area $|D|$ of D is

$$|D| = \iint\limits_D dx\,dy,$$

and the coordinates $(\bar{x}, \bar{y})$ of the center of gravity of D are

$$\bar{x} = \frac{1}{|D|} \iint\limits_D x\,dx\,dy, \qquad \bar{y} = \frac{1}{|D|} \iint\limits_D y\,dx\,dy.$$

Without loss of generality suppose that the line passing through the center of gravity of D is the x-axis. Then we have

$$\iint\limits_D y\,dx\,dy = 0. \tag{4.13}$$

Denote by D^+ the part of D above the x-axis and by D^- the other part of D. Since the roles of D^+ and D^- are interchangeable, it suffices to show that

$$\frac{|D^+|}{|D^-|} \geq \frac{4}{5}$$

or

$$\left| \iint\limits_{D^+} dx\,dy \right| \geq \frac{4}{5} \left| \iint\limits_{D^-} dx\,dy \right|, \tag{4.14}$$

under the assumption (4.13), i.e.

$$\iint\limits_{D^+} y\, dx\, dy \; = \; -\iint\limits_{D^-} y\, dx\, dy.$$

That is to say, the relation (4.14) holds if D^+ and D^- have the same moment with respect to x-axis.

The <u>proof</u> of Theorem 4.1.

1. Symmetrization. Suppose D is given by

$$c \le y \le d, \qquad \phi_1(y) \le x \le \phi_2(y).$$

Then

$$\iint\limits_{D} dx\, dy \; = \; \int_c^d \int_{\phi_1(y)}^{\phi_2(y)} dx\, dy \; = \; \int_c^d \Big(\phi_2(y) - \phi_1(y)\Big)\, dy$$

$$= \int_c^d \int_{-\frac{1}{2}(\phi_2(y) - \phi_1(y))}^{\frac{1}{2}(\phi_2(y) - \phi_1(y))} dx\, dy$$

and

$$\iint\limits_{D} y\, dx\, dy \; = \; \int_c^d y \int_{-\frac{1}{2}(\phi_2(y) - \phi_1(y))}^{\frac{1}{2}(\phi_2(y) - \phi_1(y))} dx\, dy.$$

Let D' denote the region

$$c \le y \le d, \qquad -\tfrac{1}{2}\,(\phi_2(y) - \phi_1(y)) \le x \le \tfrac{1}{2}(\phi_2(y) - \phi_1(y)).$$

We proceed to prove that D' is also a convex body. In fact, if $(x_1, y_1) \in D'$ and $(x_2, y_2) \in D'$, then

$$-\tfrac{1}{2}(\phi_2(y_1) - \phi_1(y_1)) \le x_1 \le \tfrac{1}{2}(\phi_2(y_1) - \phi_1(y_1))$$

$$-\tfrac{1}{2}(\phi_2(y_2) - \phi_1(y_2)) \le x_2 \le \tfrac{1}{2}(\phi_2(y_2) - \phi_1(y_2)).$$

Let

$$x = x_1 + t(x_2 - x_1), \quad y = y_1 + t(y_2 - y_1), \qquad 0 \le t \le 1.$$

Then

$$-\tfrac{1}{2}((1-t)(\phi_2(y_1) - \phi_1(y_1)) + t(\phi_2(y_2) - \phi_1(y_2))) \le x$$

$$\le \tfrac{1}{2}((1-t)(\phi_2(y_1) - \phi_1(y_1)) + t(\phi_2(y_2) - \phi_1(y_2))).$$

Since D is convex, i.e.

$$(1-t)\phi_1(y_1) + t\phi_1(y_2) \ge \phi_1(y)$$

and

$$(1-t)\phi_2(y_1) + t\phi_2(y_2) \le \phi_2(y),$$

it follows that $(x, y) \in D'$. This means that D' can be used instead of D. Hence without loss of generality, we may suppose that D is a convex region symmetrical with respect to the y-axis.

2. **Straightening.** Let P and Q be the intersection points of the x-axis and the boundary of D. It is evident that a point R on the y-axis can be determined such that D^+ and triangle $\triangle$ PQR have the same moment with respect to the x-axis, i.e.

$$\iint_{D^+} y \, dx \, dy = \iint_{\triangle PQR} y \, dx \, dy.$$

We proceed to prove that

$$|D^+| = \iint_{D^+} dx\,dy \geq \iint_{\triangle PQR} dx\,dy.$$

In fact, let (x_0, y_0) and $(-x_0, y_0)$ be the intersection points of $\triangle PQR$ and D^+ (see Figure 4.12). Then we have $y > y_0$ if $(x,y) \in \triangle PQR - D^+$, and $y < y_0$ if $(x,y) \in D^+ - \triangle PQR$. The assertion follows.

Prolong the edges RP and RQ to RP' and RQ' so that $P'Q'$ is parallel to the x-axis and so that D^- and $PQQ'P'$ have the same moment about the x-axis, i.e.

$$\iint_{D^-} y\,dx\,dy = \iint_{PQP'Q'} y\,dx\,dy.$$

It follows that

$$|D^-| = \iint_{D^-} dx\,dy \leq \iint_{PQP'Q'} dx\,dy = |PQP'Q'|.$$

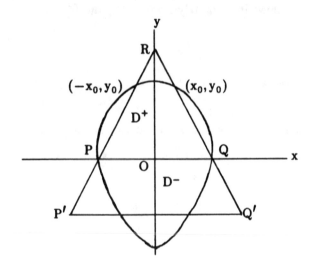

Figure 4.12

Thus

$$\frac{|D^+|}{|D^-|} \geq \frac{|\triangle PQR|}{|PQP'Q'|}. \tag{4.15}$$

Since

$$\iint_{\triangle PQR} y\,dx\,dy = -\iint_{PQP'Q'} y\,dx\,dy,$$

the center of gravity of $\triangle P'Q'R$ is still the origin 0.

3. It follows that $|RO|$ equals $\frac{2}{3}$ times the length of the altitude on $P'Q'$ in $\triangle P'Q'R$ and we have

$$|\triangle PQR| = \tfrac{4}{9}|\triangle P'Q'R|$$

and therefore

$$\frac{|\triangle PQR|}{|PQP'Q'|} = \tfrac{4}{5}. \tag{4.16}$$

The theorem follows immediately from (4.15) and (4.16). ☐

§ 4.11. Qie Kuai Fa.

In this section we will introduce an algorithm to find the maximum point of the function $z = f(x_1, \ldots, x_s)$ defined on a s-dimensional region R_0. Take a point $P_0^0(x_1^0, \ldots, x_s^0)$ of R and take s neighboring points $P_0^i(x_1^0, \ldots, x_{i-1}^0, x_i^0 + \delta_i, x_{i+1}^0, \ldots, x_s^0)$, $1 \leq i \leq s$, where, for each i, δ_i is sufficiently small so that P_0^i belongs to R. Arrange the first $s+1$ trials at P_0^i ($0 \leq i \leq s$), and denote the corresponding experimental results by $f(P_0^i)$ ($0 \leq i \leq s$). Let P_0' be the point

$$(x_1^0 + \frac{f(P_0^1) - f(P_0^0)}{\delta_1}, \dots, x_s^0 + \frac{f(P_0^s) - f(P_0^0)}{\delta_s}).$$

Draw a hyperplane π_0 through P_0^0, perpendicular to the line connecting $P_0{}'$ and P_0^0. Then π_0 divides R_0 into two parts. Eliminate the part without the point $P_0{}'$, keep the other part and denote it by R_1. Take a point P_1^0 of R_1 and then repeat the above process until the maximum point (or a nearby point) is found. This method is called the method of cutting the trial region piece by piece (or Qie Kuai Fa in Chinese).

The idea behind Qie Kuai Fa is reasonable, since π_0 is the tangent plane to the isoplethic surface at P_0^0 and the remaining region contains the maximum point of the function. Therefore the main problem is how to choose the points P_0^0, P_1^0, ..., so that the maximum point of z will be attained as quickly as possible. There are many suggestions concerning the choice of P_0^0, P_1^0, ... For example, when R is a convex body it is sufficient to take P_0^0 to be the center of gravity of R_0, since any hyperplane passing through the center of gravity of R_0 divides R_0 into two parts, and the volume of the remaining region R_1 is always less than

$$1 - \frac{1}{e}$$

times the volume of the original region R_0 (see Theorem 4.1). Notice that the number $1 - \frac{1}{e}$ is independent of the dimension s. Hence if one wants to obtain a region containing the maximum point of z and with volume $\leq \epsilon$, then the number of necessary trials is of order of magnitude $s \ln \frac{1}{\epsilon}$.

Since the coordinates of the center of gravity of a body can be represented by multiple integrals over the body, they can be obtained using the numerical integration method in s-dimensional space described in Chapter 5.

§ 4.12 The 0-1 variable method.

Before optimum seeking methods became a topic in applied mathematics, there were already some methods to find the maximum point of a function. Suppose that a parameter can only take values at certain levels. In the simplest case there are only two choices for a parameter and then it can be represented by a 0-1 variable. For instance, if the temperature can only assume the values 300°C and 600°C, then introduce a new variable that takes the value 1 if the temperature is 300°C and 0 otherwise. More precisely, define the 0-1 variable η by

$$\eta = \frac{600°C - \xi}{600°C - 300°C},$$

where the parameter ξ represents the temperature.

Now we will introduce the 0-1 variable method with modifications.

Suppose that a parameter ξ can only take values from among n possible levels

$$\xi_1, \xi_2, \ldots, \xi_n.$$

at each choice. Then it may be represented by 0-1 variables $x_1, \ldots, x_n$, where each x_i takes only values 0 or 1 and $x_1 + \ldots + x_n = 1$.

Let T be an objective function depending on k parameters $\xi^1, \ldots, \xi^k$, where now parameter ξ^i has n_i possible levels

$$\xi_1^i, \ldots, \xi_{n_i}^i,$$

and corresponding 0-1 variables $x_1^i, \ldots, x_{n_i}^i$. It is often assumed that T is a linear model (but the reliability of this assumption is quite small).

$$T = \sum_{i=1}^{k} \sum_{j=1}^{n_i} \alpha_j^i x_j^i, \tag{4.17}$$

where, for all $1 \leq i \leq k$,

$$\sum_{j=1}^{n_i} x_j^i = 1, \qquad x_j^i = 0 \text{ or } 1.$$

Substituting

$$x_{n_i}^i = 1 - \sum_{j=1}^{n_i-1} x_j^i,$$

into (4.17), we have

$$T = \sum_{i=1}^{k} \left(\sum_{j=1}^{n_i-1} \alpha_j^i x_j^i + \alpha_{n_i}^i \left(1 - \sum_{j=1}^{n_i-1} x_j^i \right) \right)$$

$$= \sum_{i=1}^{k} \alpha_{n_i}^i + \sum_{i=1}^{k} \sum_{j=1}^{n_i-1} (\alpha_j^i - \alpha_{n_i}^i) x_j^i. \tag{4.18}$$

The $n_1 + \ldots + n_k - k$ independent 0-1 variables x_j^i in (4.18) satisfy

$$\sum_{j=1}^{n_i-1} x_j^i \leq 1, \qquad 1 \leq i \leq k.$$

Arrange experiments such that each of the k parameters $\xi^1, \ldots, \xi^k$ assumes a certain level, i.e. take a set of 0-1 variables x_j^i. Then we observe a corresponding value of T. After M different sets of experiments we have a system of M linear equations whose unknowns are

$$\sum_{i=1}^{k} \alpha_{n_i}^i, \quad \alpha_j^i - \alpha_{n_i}^i \quad (1 \leq i \leq k, \ 1 \leq j \leq n_i - 1). \tag{4.19}$$

The total number of unknowns is

$$N = n_1 + \ldots + n_k - k + 1.$$

Hence if $M = N$, then by solving the system of linear equations, we obtain unique values for the unknowns in (4.19), and therefore the objective function T is uniquely determined. Notice that the experiments should be arranged so that the determinant of the coefficients of the equations is not equal to zero.

Once the unknowns have been determined, the best choice for the level of ξ^i can be obtained as follows. Define $j_0(i) = n_i$ and $x^i_{n_i} = 1$, i.e. ξ^i takes its n_i-th level, if

$$\alpha^i_j - \alpha^i_{n_i} < 0, \quad 1 \le j \le n_i - 1.$$

Otherwise let $j_0(i)$ be such that $\alpha^i_{j_0(i)} - \alpha^i_{n_i}$ is a maximum among $\alpha^i_j - \alpha^i_{n_i}$, $1 \le j \le n_i - 1$, and then take $x^i_{j_0(i)} = 1$.

According to this model, T attains its maximum at

$$\xi^1 = \xi^1_{j_0(1)}, \ldots, \xi^k = \xi^k_{j_0(k)},$$

and then the conclusion is made that this is the maximum point of T. As a matter of fact, the model is evidently not reliable, since there exists no maximum point for the linear function in general. Hence it is only a maximum point in some specified sense. That is to say the above method does not belong to the category of optimum seeking methods for which one can estimate the rate of convergence.

Since the assumption of a linear model is often unreliable, we do not discuss in detail the problem of how to choose the x^i_j's, nor the problem of how to analyse the experimental results by mathematical statistics. In conclusion, this is a method which can not ensure that the optimum solution is obtained, although it is optimal among the given $n_1 n_2 \ldots n_k$ points.

Example. Consider the problem of finding the best combination of three kinds of chemical reagents, where for each reagent there are three quantity choices, and after a combination is made, the mix will be heated at one of three possible temperature levels.

This is a problem with four parameters where the first three denote the quantities of the chemical reagents and the last one represents the temperature. By the above method, a linear model for this problem can be determined with

$$3 \times 4 - 4 + 1 = 9$$

experiments.

References.

Hua Loo Keng. Theory of Optimization. Science Press, Beijing, 1981.

Hua Loo Keng and Wang Yuan. Applications of Number Theory to Numerical Analysis. Science Press, Beijing, 1978 and Springer Verlag, 1981.

Editor's note: See also the references at the end of Chapter 3.

CHAPTER 5

THE GOLDEN NUMBER AND NUMERICAL INTEGRATION

§ 5.1 Introduction.

The golden number $\omega = \dfrac{\sqrt{5}-1}{2}$ is not only useful for the golden section method, but it also plays an important role in the theory of diophantine approximation. This inspired us to think about the connection between numerical integration and the golden number. In 1960 we found the very efficient quadrature formula

$$\int_0^1 \int_0^1 f(x, y)\, dx\, dy \approx \frac{1}{F_m} \sum_{k=1}^{F_m} f\left(\frac{k}{F_m}, \left\{\frac{F_{m-1}}{F_m} k\right\}\right), \qquad (5.1)$$

that is based on the rational approximation of ω

$$\left| \frac{F_{m-1}}{F_m} - \omega \right| < \frac{1}{\sqrt{5}\, F_m^2},$$

and where $\left\{ \xi \right\}$ denotes the fractional part of ξ.

In order to extend the method to the s-dimensional case where s > 2, it suffices to generalize the Fibonacci numbers associated with the golden number ω.

To begin, since $\omega = 2 \cos \frac{2\pi}{5}$, the golden number is generalized to the set

$$\omega_j = 2 \cos \frac{2\pi j}{m}, \qquad 1 \le j \le s-1,$$

where m denotes an integer ≥ 5 and $s = \dfrac{\phi(m)}{2}$ (ϕ is the Euler

111

function). The set $\{\,1,\ \omega_j\ (\,1 \leq j \leq s-1\,)\}$ is an integral basis of the cyclotomic field $\mathcal{R}_s = \mathbb{Q}(\cos\frac{2\pi}{m})$. By using a set of independent units of $\mathcal{R}_s$, one may obtain a sequence of sets of integers

$$(h_{1,l}, \ldots, h_{s-1,l}, n_l) \qquad (\, l = 1, 2, \ldots) \qquad\qquad (5.2)$$

such that

$$\left| \frac{h_{j,l}}{n_l} - \omega_j \right| < c(\mathcal{R}_s)\, n_l^{-1-\frac{1}{s-1}}, \qquad 1 \leq j \leq s-1,$$

where $c(\mathcal{R}_s)$ denotes a positive constant that depends only on $\mathcal{R}_s$. The sequence (5.2) is regarded as the generalization of the Fibonacci sequence. We now obtain a s-dimensional quadrature formula that involves only a single summation

$$\int_0^1 \cdots \int_0^1 f(x_1, \ldots, x_s)\, dx_1 \ldots dx_s \approx$$

$$\approx \frac{1}{n_l} \sum_{k=1}^{n_l} f(\frac{k}{n_l}, \left\{\frac{h_{1,l}\,k}{n_l}\right\}, \ldots, \left\{\frac{h_{s-1,l}\,k}{n_l}\right\}). \qquad (5.3)$$

This formula is very efficient in practice. However, since the method must be used on computers and since it is often used by scientists with some mathematical training, it is certainly not needed for popularization. Besides, a table containing certain sets of integers (5.2) with error estimations for the corresponding quadrature formulas (5.3) for the cases $2 \leq s \leq 18$ was compiled for practical use.

The sequence of sets (5.2) can be used not only in numerical integration, but also in situations where uniformly distributed pseudo random numbers are being employed, in particular it may be used to find a starting point for some optimum seeking methods of several variables.

Our research on numerical integration in multi-dimensional space is an example that shows that our involvement in popularizing mathematical methods in fact often stimulates our theoretical research rather than hinders it.

In this chapter we only give an estimate of the error term for the formula (5.1). For a full exposition of the theory we refer the reader to our book "Applications of Number Theory to Numerical Analysis" (Hua Loo Keng and Wang Yuan [1982]).

§ 5.2 Lemmas.

Let $f(x, y)$ be a periodic function with period 1 for each variable, and whose Fourier expansion is given by

$$f(x, y) = \sum_{m=-\infty}^{\infty} \sum_{n=-\infty}^{\infty} C(m, n)\, e^{2\pi i\,(mx + ny)},$$

where

$$|\, C(m, n)\, | \le C\left(\overline{m}\,\overline{n}\right)^{-\alpha},$$

in which $\overline{x} = \text{MAX}\,(1, |x|)$, and $\alpha\ (> 1)$ and $C\ (> 0)$ are constants. The class of these functions is denoted by $E(\alpha, C)$.

Lemma 5.1. Let m and $n\ (> 0)$ be integers. Then

$$\frac{1}{n} \sum_{k=0}^{n-1} e^{2\pi i\,\frac{mk}{n}} = \begin{cases} 1, & \text{if } n \mid m, \\ 0, & \text{otherwise.} \end{cases}$$

Proof. The lemma is obviously true if $n \mid m$. Otherwise, we have that

$$e^{2\pi i \frac{m}{n}} \neq 1, \text{ and so}$$

$$\sum_{k=0}^{n-1} e^{2\pi i \frac{mk}{n}} = (e^{2\pi i \frac{mn}{n}} - 1)(e^{2\pi i \frac{m}{n}} - 1)^{-1} = 0. \qquad \square$$

Lemma 5.2. For all positive integers n, a_1 and a_2 with $(a_i, n) = 1$, $i = 1, 2$, (where $(\ ,\ )$ denotes greatest common divisor), we have

$$\operatorname*{Sup}_{f \in E(\alpha, C)} \left| \int_0^1 \int_0^1 f(x, y)\, dx\, dy - \frac{1}{n} \sum_{k=1}^{n} f(\frac{a_1 k}{n}, \frac{a_2 k}{n}) \right| \leq$$

$$\leq C \left(\wedge(a_1, a_2) + 2^{\alpha+1}\left(1 + 2\,\zeta(\alpha)\right)^2 n^{-\alpha} \right),$$

where $\zeta(\alpha) = \displaystyle\sum_{m=1}^{\infty} m^{-\alpha}$ and

$$\wedge(a_1, a_2) = \sideset{}{'}\sum_{\substack{a_1 m_1 + a_2 m_2 \equiv 0 (\mathrm{mod}\ n) \\ -\frac{n}{2} < m_i \leq \frac{n}{2}}} \left(\overline{m_1}\ \overline{m_2}\right)^{-\alpha},$$

in which $\displaystyle\sideset{}{'}\sum$ denotes a sum with $m_1 = m_2 = 0$ neglected.

Proof. It follows from Lemma 5.1 that

$$\frac{1}{n} \sum_{k=1}^{n} f(\frac{a_1 k}{n}, \frac{a_2 k}{n}) =$$

$$= \frac{1}{n} \sum_{k=1}^{n} \sum_{m_1=-\infty}^{\infty} \sum_{m_2=-\infty}^{\infty} C(m_1, m_2)\, e^{2\pi i \frac{(a_1 m_1 + a_2 m_2)\, k}{n}}$$

$$= \sum_{m_1=-\infty}^{\infty} \sum_{m_2=-\infty}^{\infty} C(m_1, m_2)\, \frac{1}{n} \sum_{k=1}^{n} e^{2\pi i \frac{(a_1 m_1 + a_2 m_2)\, k}{n}}$$

$$= \sum_{a_1 m_1 + a_2 m_2 \equiv 0 (\text{mod } n)} C(m_1, m_2)$$

$$= C(0, 0) + \sideset{}{'}\sum_{a_1 m_1 + a_2 m_2 \equiv 0 (\text{mod } n)} C(m_1, m_2).$$

Since

$$C(0, 0) = \int_0^1 \int_0^1 f(x, y) \, dx \, dy,$$

we have

$$\operatorname*{Sup}_{f \in E(\alpha, C)} \left| \int_0^1 \int_0^1 f(x, y) \, dx \, dy - \frac{1}{n} \sum_{k=1}^n f(\frac{a_1 k}{n}, \frac{a_2 k}{n}) \right| \le C \, \Omega(a_1, a_2),$$

where

$$\Omega(a_1, a_2) = \sideset{}{'}\sum_{a_1 m_1 + a_2 m_2 \equiv 0 (\text{mod } n)} \left(\overline{m_1} \; \overline{m_2} \right)^{-\alpha}.$$

If $(m_1^{(0)}, m_2^{(0)})$ is a solution of the congruence

$$a_1 m_1 + a_2 m_2 \equiv 0 \; (\text{mod } n), \tag{5.4}$$

where $-\frac{n}{2} < m_i^{(0)} \le \frac{n}{2}$ (i = 1, 2), then

$$m_i = m_i^{(0)} + l_i \, n, \qquad i = 1, 2, \tag{5.5}$$

is also a solution of (5.4). On the other hand, any solution of (5.4) can be represented in the form (5.5), and therefore

$$\Omega(a_1, a_2) =$$

$$= \sum_{\substack{a_1 m_1^{(0)} + a_2 m_2^{(0)} \equiv 0 (\text{mod } n) \\ -\frac{n}{2} < m_i^{(0)} \leq \frac{n}{2}}}' \left(\overline{(m_1^{(0)} + l_1 n)} \, \overline{(m_2^{(0)} + l_2 n)} \right)^{-\alpha}.$$

Since $(a_i, n) = 1$ $(i = 1, 2)$, it follows that for each value of $m_1^{(0)}$, $m_2^{(0)}$ is uniquely determined, and vice versa. Since

$$\sum_{-\frac{n}{2} < m \leq \frac{n}{2}} \sum_{l=-\infty}^{\infty} \overline{(m + ln)}^{-\alpha} \leq 1 + 2 \sum_{m=1}^{\infty} \overline{m}^{-\alpha} = 1 + 2\zeta(\alpha)$$

and

$$\sum_{l=-\infty}^{\infty}{}' \overline{(m + ln)}^{-\alpha} \leq 2 \sum_{l=1}^{\infty} \left(n(1 - \tfrac{1}{2}) \right)^{-\alpha} \leq 2^{\alpha+1} \zeta(\alpha) \, n^{-\alpha},$$

we have

$$\Omega(a_1, a_2) - \Lambda(a_1, a_2) \leq$$

$$2 \sum_{\substack{a_1 m_1^{(0)} + a_2 m_2^{(0)} \equiv 0 (\text{mod } n) \\ -\frac{n}{2} < m_i^{(0)} \leq \frac{n}{2}}} \sum_{l_1 = -\infty}^{\infty}{}' \overline{(m_1^{(0)} + l_1 n)}^{-\alpha} \sum_{l_2 = -\infty}^{\infty} \overline{(m_2^{(0)} + l_2 n)}^{-\alpha}$$

$$\leq 2^{\alpha+1} \left(1 + 2\zeta(\alpha) \right)^2 n^{-\alpha}. \qquad \square$$

§ 5.3 Error estimation for the quadrature formula.

<u>Theorem</u> 5.1. Let n be an integer > 3. Then

$$\sup_{f \in E(\alpha, C)} \left| \int_0^1 \int_0^1 f(x, y) \, dx \, dy - \frac{1}{F_n} \sum_{k=1}^{F_n} f\left(\frac{k}{F_n}, \frac{F_{n-1} k}{F_n}\right) \right| \le$$

$$\le C \left(4 \, \zeta(\alpha) \, 5^\alpha \, \omega^{-\alpha} \, (\ln \tfrac{1}{\omega})^{-1} \, F_n^{-\alpha} \, \ln (\sqrt{5} \, \omega \, F_n) + \right.$$

$$\left. + 2^{\alpha+1} \left(1 + 2 \, \zeta(\alpha)\right)^2 F_n^{-\alpha} \right).$$

Proof. We start to estimate the sum

$$\wedge(1, F_{n-1}) = \underset{\substack{m_1 + F_{n-1} m_2 \equiv 0 (\text{mod } F_n) \\ -\frac{F_n}{2} < m_i \le \frac{F_n}{2}}}{\sum{}'} \left(\overline{m_1} \, \overline{m_2}\right)^{-\alpha}.$$

It is evident that $m_2 \ne 0$, otherwise it follows from $F_n \mid m_1$ and $-\frac{F_n}{2} < m_1 \le \frac{F_n}{2}$ that $m_1 = 0$, which gives a contradiction. For any given m_2, we have that $m_1 = y \, F_n - F_{n-1} m_2$, where y is an integer satisfying

$$-\tfrac{1}{2} < y - \frac{F_{n-1}}{F_n} m_2 \le \tfrac{1}{2}.$$

Hence

$$\wedge(1, F_{n-1}) \le 2 \sum_{1 \le x \le \frac{F_n}{2}} \left(x \, \overline{(F_{n-1} x - F_n y)}\right)^{-\alpha} \le 2 \sum_{m=2}^{n-1} J_m, \qquad (5.6)$$

where

$$J_m = \sum_{F_{m-1} \leq x < F_m} \left(x \overline{(F_{n-1} x - F_n y)} \right)^{-\alpha}.$$

Notice that $F_{n-1} \geq \frac{1}{2}(F_{n-1} + F_{n-2}) = \frac{1}{2} F_n$. By Theorem 2.2, we have

$$F_m F_{m-2} - F_{m-1}^2 = (-1)^{m-1},$$

and therefore for any given integers x and y, the system of equations

$$x = F_{m-1} u + F_m v,$$

$$y = F_{m-2} u + F_{m-1} v,$$

has a unique integral solution u, v. It is evident that

$$u v < 0$$

if $F_{m-1} \leq x < F_m$. Now we shall prove that for any two distinct integers x and x' in the interval $[F_{m-1}, F_m)$, the corresponding u and u' are also distinct. Otherwise, from

$$x = F_{m-1} u + F_m v,$$

$$x' = F_{m-1} u + F_m v',$$

we have $F_m \mid (x - x')$ which is impossible, and therefore the assertion follows. From Lemma 3.1 we have

$$F_{n-1} F_k - F_n F_{k-1} = (-1)^{k-1} F_{n-k},$$

and so

$$|F_{n-1} x - F_n y| =$$

$$= |F_{n-1}(F_{m-1} u + F_m v) - F_n(F_{m-2} u + F_{m-1} v)|$$

$$= |(F_{n-1}F_{m-1} - F_n F_{m-2}) u + (F_{n-1}F_m - F_n F_{m-1}) v|$$

$$= |F_{n-m+1} u - F_{n-m} v| \geq F_{n-m+1} |u|.$$

Therefore

$$J_m \leq \sum_{F_{m-1} \leq x < F_m} \left(x F_{n-m+1} |u| \right)^{-\alpha} \left(F_{m-1} F_{n-m+1} \right)^{-\alpha} \sum_{u=-\infty}^{\infty}{}' \bar{u}^{-\alpha}$$

$$= 2 \zeta(\alpha) \left(F_{m-1} F_{n-m+1} \right)^{-\alpha}. \tag{5.7}$$

It follows from Lemma 3.1 that for $k \geq 1$,

$$\omega^{-k+1} \geq F_k \geq \frac{1}{\sqrt{5}} \omega^{-k+1} (\omega^{-1} - \omega) = \frac{1}{\sqrt{5}} \omega^{-k+1},$$

and so

$$k-2 \leq \frac{\ln F_n + \ln (\sqrt{5}\,\omega)}{\ln (\omega^{-1})}$$

and

$$F_{m-1} F_{n-m+1} \geq \frac{1}{5} \omega^{-n+2}.$$

Hence, by (5.7), we have

$$J_m \le 2\,\zeta(\alpha)\,5^\alpha\,\omega^{-2\alpha}\,\omega^{n\alpha} \le 2\,\zeta(\alpha)\,5^\alpha\,\omega^{-\alpha}\,F_n^{-\alpha}$$

and therefore, by (5.6),

$$\wedge(1,\,F_{n-1}) \le 4\,\zeta(\alpha)\,5^\alpha\omega^{-\alpha}\,(n-2)\,F_n^{-\alpha} \le$$

$$\le 4\,\zeta(\alpha)\,5^\alpha\,\omega^{-\alpha}\,(\ln\tfrac{1}{\omega})^{-1}\,(\ln F_n + \ln(\sqrt5\,\omega))\,F_n^{-\alpha}.$$

The theorem follows by Lemma 5.2. $\square$

§ 5.4 A result for Ω and a lower bound for the quadrature formula.

<u>Theorem</u> 5.2. For any given integer $n > 1$ and two integers a_1 and a_2 that are relatively prime with n, we have

$$\underset{f\in E(\alpha,\,C)}{\text{Sup}}\left|\int_0^1\int_0^1 f(x,\,y)\,dx\,dy - \frac{1}{n}\sum_{k=1}^n f\left(\frac{a_1 k}{n},\,\frac{a_2 k}{n}\right)\right| \ge \alpha\,C\,\frac{\ln n}{n^\alpha}.$$

Proof. Consider the function

$$f(x,\,y) = C\sum_{m_1=-\infty}^\infty\sum_{m_2=-\infty}^\infty \left(\overline{m_1}\,\overline{m_2}\right)^{-\alpha} e^{2\pi i(m_1 x + m_2 y)}$$

in $E(\alpha,\,C)$. We have

$$\left|\int_0^1\int_0^1 f(x,\,y)\,dx\,dy - \frac{1}{n}\sum_{k=1}^n f\left(\frac{a_1 k}{n},\,\frac{a_2 k}{n}\right)\right| = C\,\Omega,$$

where

$$\Omega = \underset{a_1 m_1 + a_2 m_2 \equiv 0(\text{mod } n)}{{\sum}'}\left(\overline{m_1}\,\overline{m_2}\right)^{-\alpha}.$$

Since $(a_1, n) = 1$, there exists an integer $\bar{a}_1$ such that

$$\bar{a}_1 \, a_2 \equiv 1 \pmod{n}.$$

Thus

$$\Omega = \sideset{}{'}\sum_{m_1 + a m_2 \equiv 0 \pmod{n}} \left(\overline{m}_1 \, \overline{m}_2 \right)^{-\alpha},$$

where

$$a \equiv \bar{a}_1 \, a_2 \pmod{n}, \qquad 1 \le a < n.$$

Expand $\frac{a}{n}$ into the simple continued fraction and suppose that

$$\frac{p_0}{q_0}, \; \frac{p_1}{q_1}, \; \ldots, \; \frac{p_m}{q_m}$$

are its convergents, where $p_m = a$ and $q_m = n$. By Theorem 2.5, we have

$$\left| \frac{p_l}{q_l} - \frac{p_m}{q_m} \right| \le \frac{1}{q_l \, q_{l+1}}, \qquad 0 \le l \le m-1,$$

i.e.

$$| p_l \, q_m - q_l \, p_m | \le \frac{q_m}{q_{l+1}}, \qquad 0 \le l \le m-1.$$

Since $m_1 = p_l \, q_m - q_l \, p_m$ and $m_2 = q_l$ $(0 \le l \le m-1)$ are solutions of the congruence

$$m_1 + a \, m_2 \equiv 0 \pmod{n},$$

it follows that

$$\Omega \ge \sum_{l=0}^{m-1} \left(\overline{q_l \, (p_l \, q_m - q_l \, p_m)} \right)^{-\alpha} \ge \sum_{l=0}^{m-1} (q_m \, q_l)^{-\alpha} \, q_{l+1}^{\alpha}$$

$$\geq m \left(\frac{q_m}{q_{m-1}} \frac{q_{m-1}}{q_{m-2}} \cdots \frac{q_1}{q_0} \right)^{\frac{\alpha}{m}} q_m^{-\alpha} = m \, q_m^{\frac{\alpha}{m}} \, q_m^{-\alpha}$$

$$= (m \, e^{\frac{\alpha}{m} \ln q_m}) \, q_m^{-\alpha} \geq \alpha \, q_m^{-\alpha} \ln q_m,$$

using the fact that for a finite set of positive integers the arithmetic mean is no less than the geometric mean. □

From Theorem 5.2, it follows that the error estimate given in Theorem 5.1 is, except for a constant, the best possible.

§ 5.5 Remarks.

For the trapezoid formula with $s = 2$, we have

Theorem 5.3. Let $n = m^2$. Then

$$\underset{f \in E(\alpha, C)}{\text{Sup}} \left| \int_0^1 \int_0^1 f(x, y) \, dx \, dy - \frac{1}{n} \sum_{l_1=0}^{m-1} \sum_{l_2=0}^{m-1} f\left(\frac{l_1}{m}, \frac{l_2}{m} \right) \right| \leq$$

$$\leq C \left(2 \, \zeta(\alpha) + 1 \right)^2 n^{-\frac{\alpha}{2}}. \tag{5.8}$$

Proof. By Lemma 5.1, we have

$$\frac{1}{n} \sum_{l_1=0}^{m-1} \sum_{l_2=0}^{m-1} f\left(\frac{l_1}{m}, \frac{l_2}{m} \right) =$$

$$= \frac{1}{n} \sum_{l_1=0}^{m-1} \sum_{l_2=0}^{m-1} \sum_{m_1=-\infty}^{\infty} \sum_{m_2=-\infty}^{\infty} C(m_1, m_2) \, e^{2\pi i \frac{(l_1 m_1 + l_2 m_2)}{m}}$$

$$= \sum_{m_1=-\infty}^{\infty} \sum_{m_2=-\infty}^{\infty} C(m_1, m_2) \frac{1}{n} \sum_{l_1=0}^{m-1} \sum_{l_2=0}^{m-1} e^{2\pi i \frac{(l_1 m_1 + l_2 m_2)}{m}}$$

$$= C(0, 0) + \sideset{}{'}\sum_{\substack{m_1 \equiv 0 \pmod{m} \\ m_2 \equiv 0 \pmod{m}}} C(m_1, m_2),$$

and therefore

$$\underset{f \in E(\alpha, C)}{\text{Sup}} \left| \int_0^1 \int_0^1 f(x, y) \, dx \, dy - \frac{1}{n} \sum_{l_1=0}^{m-1} \sum_{l_2=0}^{m-1} f\left(\frac{l_1}{m}, \frac{l_2}{m}\right) \right|$$

$$\leq C \sideset{}{'}\sum_{\substack{m_1 \equiv 0 \pmod{m} \\ m_2 \equiv 0 \pmod{m}}} \left(\overline{m_1} \, \overline{m_2}\right)^{-\alpha} \leq C \, m^{-\alpha} \left(\sum_{k=-\infty}^{\infty} \overline{k}^{-\alpha}\right)^2$$

$$= C \left(2 \zeta(\alpha) + 1\right)^2 n^{-\frac{\alpha}{2}}. \qquad\qquad \Box$$

Let $f(x, y) = C \, m^{-\alpha} \left(e^{2\pi i m x} + e^{-2\pi i m x}\right)$. Then $f(x, y) \in E(\alpha, C)$ and

$$\left| \int_0^1 \int_0^1 f(x, y) \, dx \, dy - \frac{1}{n} \sum_{l_1=0}^{m-1} \sum_{l_2=0}^{m-1} f\left(\frac{l_1}{m}, \frac{l_2}{m}\right) \right| = 2 \, C \, n^{-\frac{\alpha}{2}}.$$

This means that the error term $n^{-\frac{\alpha}{2}}$ in (5.8) can not be improved any further, and this error term is much worser than the term $n^{-\alpha} \ln n$ given in Theorem 5.1. We also notice that the error term for the quadrature formula using the Monte Carlo method is only $O(n^{-\frac{1}{2}})$, in the probabilistic sense.

124

References.

Hua Loo Keng, and Wang Yuan. Applications of Number Theory to Numerical Analysis. Science Press, Beijing, 1978 and Springer Verlag, 1981.

Editor's note: Some additional references are the following.

Davis P.J., and P.Rabinowitz. Methods of Numerical Integration. Academic Press, 1975.

Lang S. Algebraic Number Theory. Springer Verlag, 1986.

Stewart I.N., and D.O. Tall. Algebraic Number Theory. 2nd ed., Chapman and Hall Ltd., 1987.

Stroud A.M. Numerical Quadrature and Solution of Ordinary Differential Equations. Springer Verlag, 1974.

CHAPTER 6

OVERALL PLANNING METHODS

§ 6.1 Introduction.

Overall planning methods (<u>Tong</u> <u>Chou</u> <u>Fa</u> in Chinese) are methods to deal with problems in the management and organization of production. As a result of a long period of contacts (1965 to 1982) with a large number of industries, we have found that this topic offers some of the most appropriate techniques for popularization.

CPM, PERT, the method for machine scheduling, and the graphical and the simplex method for transportation problems are methods which have been popularized in China. In particular, CPM and PERT are widely used and we shall introduce them in these two chapters.

§ 6.2 Critical Path Method.

We will illustrate this method by an example involving a project with nine activities. The project and its activities can be modeled by a directed graph. Each activity is represented by a directed arc between two nodes. Each node represents a point in time when all activities leading into that node are completed, and none of the activities leaving the node are underway. For example, $\boxed{1}\!\longrightarrow\!\boxed{2}$ represents the activity $(1\sim2)$. The time duration (say in days) of an activity will be represented by assigning a number to the corresponding arc, for instance

125

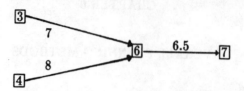 Figure 6.1 shows a situation where activity $(3\sim6)$ can be started only after activities $(3\sim6)$ and $(4\sim6)$ are completed.

Figure 6.1

Sometimes it is convenient to introduce artificial activities with zero time duration (<u>pseudo</u> activities), for example ①———⁰———→②. A typical diagram is drawn in Figure 6.2. Such a diagram is called a Critical Path diagram (CP diagram), or <u>Tong Chou Tu</u> in Chinese.

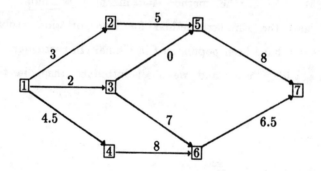

Figure 6.2

Consider the initial nodes (nodes without incoming arcs) and the end nodes (nodes without outgoing arcs) in a CP diagram. For example, there is only one initial node (①) and one end node (⑦) in Figure 6.2. For each directed path from an initial node to an end node we determine the total time duration. Note that there may be several directed paths between a given initial node and a given end node and all should be considered. For example, in Figure 6.2, the durations along the paths

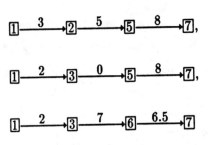

and

are $3+5+8 = 16$, $2+0+8 = 10$, $2+7+6.5 = 15.5$ and $4.5+8+6.5$ $= 19$ respectively. The total time duration of the project is equal to the largest time duration of all the paths connecting initial nodes with end nodes in the CP diagram. A path with largest duration is called a critical path (CP) or <u>Zhu</u> <u>Yao</u> <u>Mao</u> <u>Dun</u> <u>Xian</u> in Chinese, which means the principal contradiction line. Note that there may be more than one CP.

If the duration of an activity on a CP is delayed by one day, then the total duration of the project is enlarged by one day. If the duration of an activity is shortened by one day then the total duration of the project is decreased if that activity belongs to all critical paths. Since other paths may now be critical, the decrease in the total duration of the project could be less than one day. In order to avoid wasting manpower, the start of some activities that belong only to non-critical paths may be delayed for a certain amount of time.

§ 6.3 Float.

Consider the <u>early</u> <u>start</u> <u>time</u> of each activity, i.e. the time that the activity can be started if all preceding activities were started as early as possible. For instance, $(2\sim5)$ in Figure 6.2 can be started only when

$(1 \sim 2)$ is finished, hence the early start time of $(2 \sim 5)$ is 3 (the end of day 3). Similarly, the early start time for activity $(4 \sim 6)$ is 4.5. The situation is different for activity $(6 \sim 7)$. Since there are two paths from ① to ⑥, namely

$$① \xrightarrow{\ 2\ } ③ \xrightarrow{\ 7\ } ⑥ \quad \text{and} \quad ① \xrightarrow{\ 4.5\ } ④ \xrightarrow{\ 8\ } ⑥,$$

and their durations are $2+7 = 9$ and $4.5+8 = 12.5$, the early start time of activity $(6 \sim 7)$ is 12.5.

The <u>late start time</u> of an activity is the last time the activity can be started without delaying the completion of the project. Since the total duration for our example is 19 days, the project will be delayed if $(5 \sim 7)$ is not started at the beginning of the 12th day, i.e. after the 11th day has ended. Hence the late start time of $(3 \sim 6)$ is $19-8 = 11$. Similarly, the late start time for $(6 \sim 7)$ is $19-6.5 = 12.5$. Since there are two paths between ③ and ⑦,

$$③ \xrightarrow{\ 0\ } ⑤ \xrightarrow{\ 8\ } ⑦ \quad \text{and} \quad ③ \xrightarrow{\ 7\ } ⑥ \xrightarrow{\ 6.5\ } ⑦,$$

and their durations are $0+8 = 8$ and $7+6.5 = 13.5$, the late start time of $(1 \sim 3)$ is the smaller one of $19-8-2 = 9$ and $19-13.5-2 = 3.5$. Thus, in order not to delay the project, activity $(1 \sim 3)$ must be started within 3.5 days after the beginning of the project. Pictorially, the early start time and the late start time of an activity will be represented by means of an ordered pair alongside the arc, with the early start time as the first component. For example, for activity $(2 \sim 5)$ has early start time 3 and late start time $19-8-5 = 6$, and this will be represented as

$$② \underset{(3,\,6)}{\xrightarrow{\hspace{1cm}5\hspace{1cm}}} ⑤.$$

In this way we obtain the diagram in Figure 6.3.

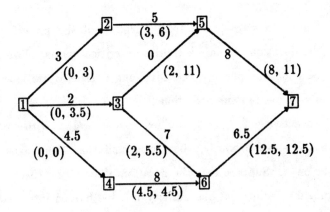

Figure 6.3

The difference between the late start time and the early start time of an activity is called the <u>float</u> (or the <u>slack</u>) of the activity. The float of each activity in the above example is shown in Table 6.1.

Table 6.1

Activity	1~2	1~3	1~4	2~5	3~5	3~6	4~6	5~7	6~7
Float	3	3.5	0	3	9	3.5	0	3	0

The float of each activity on a critical path is obviously equal to zero, and on the other hand, the critical paths may be obtained by following arcs with zero float. Assigning a specific starting time to an activity will affect the possible starting times for other activities. For example, if the float is used for (1~2), i.e. activity (1~2) is started at the beginning of day 4, then the new floats for activities (2~5) and (5~7) are both equal to zero.

130

§ 6.4 Parallel operations and overlapping operations.

In order to shorten the total duration of the project we must decrease the duration of some activities on critical paths. Two methods may be used for this purpose, one involving parallel operations and the other involving overlapping operations.

If an activity I may be divided into several activities which can be performed simultaneously, then it is said that there exist parallel operations on I. Suppose that the duration of activity (10~20) is 100 (see Figure 6.4) and that the activity can be divided into two similar activities that may be performed simultaneously.

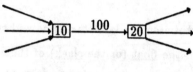

Figure 6.4

For example, 100 identical jobs on a machine are split up in two groups of 50 jobs and processed on two identical machines. Another example is the excavation of a ditch from its two ends. How does one draw the resulting CP diagram? One may introduce a pseudo activity and obtain Figure 6.5. Other possible diagrams are shown in Figures 6.6 and 6.7.

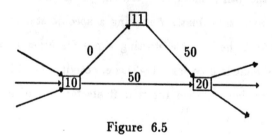

Figure 6.5

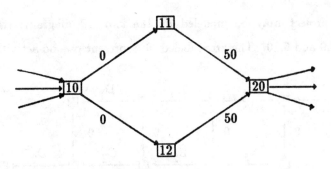

Figure 6.6

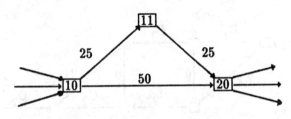

Figure 6.7

In any case, nodes ⑪ and ⑫ (or one of them) must be introduced in order to distinguish that there are now two activities between nodes ⑩ and ⑳.

Next we consider overlapping operations by means of an example. Consider the CP diagram shown in Figure 6.8, where A, B, C and D denote the names of the activities.

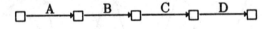

Figure 6.8

Suppose that B can be started when only a part of A is finished, and similarly C (respectively D) can be started when only a part of B (respectively C) is completed. Then A, B and C can be divided in A_1 and A_2, B_1 and B_2, and C_1 and C_2 respectively, and the project may be modeled by the two CP diagrams shown in Figures 6.9 and 6.10. The arcs labeled 0 represent pseudo activities.

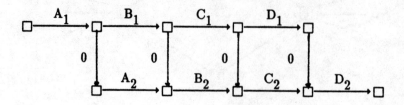

Figure 6.9

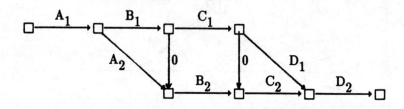

Figure 6.10

Notice that the diagram in Figure 6.11 contains a logical mistake. It implies that C_1 (resp. D_1) can be started only when A_2 (resp. B_2) is completed. This is certainly unnecessary and so the diagram is incorrect.

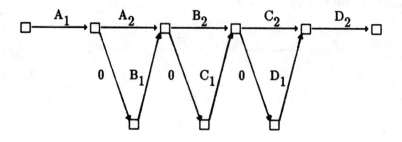

Figure 6.11

Suppose that each activity X (X = A, B, C, D) can be divided into three activities X_1, X_2 and X_3, with the following precedence requirements. For each activity X, X_i has to be completed before X_{i+1}, $i = 1, 2$. Also, activity B_i (resp. C_i, D_i) can be started only if activity A_i (resp. B_i, C_i) has been completed, for $i = 1, 2, 3$. A CP diagram for this situation is shown in Figure 6.12.

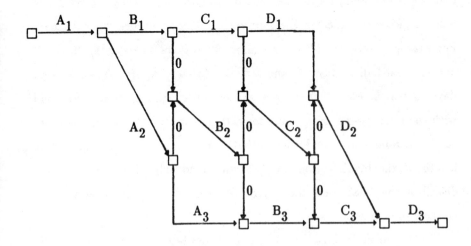

Figure 6.12

Figure 6.13 contains again a logical mistake. It says that A_3 (resp. B_3 and C_3) can be started only when B_1 (C_1 and D_1) are completed. These are unnecessary requirements.

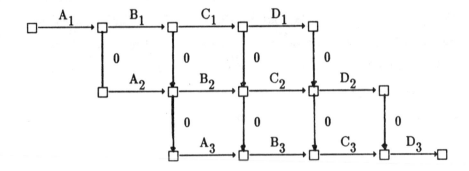

Figure 6.13

§ 6.5 Manpower scheduling

It is sometimes possible to lower the number of workers or the amount of equipment by appropriately scheduling the activities that are not part of any critical path. As an example, suppose that the manpower requirements for the example of section 6.2 are given by Table 6.2. This table indicates the number of workers needed on each day. For example, activity $(1\sim4)$ requires 3 workers on days 1, 2, 3, 4 and the first half of day 5, and activity $(3\sim6)$ requires 3 workers on days 3 to 9, etc. The regions shaded with crossing lines correspond with the manpower requirements for activities on the CP, while the other shaded regions correspond with activities not on any critical path. The length of the blank region on the right hand side of a shaded region describes the float, and the height denotes the maximum number of workers that may be used.

One must take care that the total number of workers assigned to the activities is smaller than or equal to the available manpower. When this is not the case special arrangements must be made. It is usually better to arrange the total manpower as uniformly as possible over the project. This may be accomplished by decreasing the number of workers on activities that need many workers by means of delaying the start times of some activities. For example, if the start time of activity $(5\sim7)$ is delayed by 3 days after activity $(2\sim5)$ is finished, then the number of workers in $(3\sim6)$ may be decreased to two after 4.5 days and to one after 11 days. Therefore eight workers are sufficient to complete the project within 19 days, as shown in Table 6.3. Sometimes activities need a minimum number of workers and one must pay attention to such requirements as well.

Table 6.2

Date Men	1	2	3	4	5	6	7	8	9	10	11	12	13	14	15	16	17	18	19
1																			
2		(1~4)					(4~6)									(6~7)			
3																			
4																			
5		(1~2)																	
6																			
7					(2~5)														
8																			
9									(5~7)										
10																			
11		(1~3)																	
12																			
13			(3~6)																
14																			
Total sum	6	6	8	8	8;9	9	9	9	9	6	6	6	6;7	7	7	7	5	5	5

Table 6.3

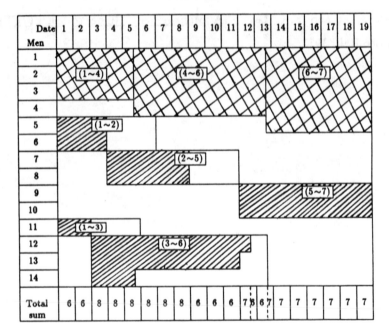

Date Men	1	2	3	4	5	6	7	8	9	10	11	12	13	14	15	16	17	18	19
1																			
2		(1~4)					(4~6)									(6~7)			
3																			
4																			
5		(1~2)																	
6																			
7					(2~5)														
8																			
9									(5~7)										
10																			
11		(1~3)																	
12			(3~6)																
13																			
14																			
Total sum	6	6	8	8	8	8	8	8	6	6	6	7;6	6;7	7	7	7	7	7	7

136

References

Hua Loo Keng. <u>Popular Lectures on Overall Planning Methods</u> (with supplements), Chinese Industry Press, Beijing, 1965.

Hua Loo Keng, Chen De Quan, Ji Lei etc. (edited). <u>A Collection of Achievements on Popularizing Optimum Seeking Methods in People's Republic of China</u>, Vol. I, II, III, IV, Science and Tech, Document Press, Beijing, 1977.

Kelley, J.E. Jr., and M.R. Walker. <u>Critical Path Planning and Scheduling</u>, Proc. of the Eastern Joint Computer Conf., Dec. 1959, 1-3.

Editor's note: Some additional references are the following.

Elmaghraby, S.E. <u>Activity Networks: Project Planning and Control by Network Models</u>, Wiley, 1977.

Moder, J.J., C.R. Phillips and E.W. Davis. <u>Project Management with CPM, PERT and Precedence Diagramming</u>, 3th ed., Van Nostrand, 1983.

Wiest, J.D., and F.K. Levy. <u>A Management Guide to PERT/CPM</u>, Prentice-Hall, 1977.

CHAPTER 7

PROGRAM EVALUATION AND REVIEW TECHNIQUE (PERT)

§ 7.1 Introduction.

Problem. Suppose a project has n activities and that three time
parameters are associated with each activity, the "optimistic duration",
the "most probable duration" and the "pessimistic duration."   The
exact duration of each activity is not known. How can we estimate the
probability that the project is completed within a certain amount of
time?

In order to make a distinction, the problem introduced in Chapter
6 is called the problem of deterministic type, while the above problem is
said to be of undeterministic type. If we are familiar with every activity
of a project so that we can estimate exactly the duration of each activity
from previously accumulated data, it is better to consider the problem as
a deterministic one. Otherwise, if a project contains an activity whose
duration can not be estimated based on previous experiences, it should be
considered as a problem of undeterministic type.

Let a, c and b denote the optimistic duration, the most
probable duration and the pessimistic duration of an activity. We use

$$\frac{a + 4c + b}{6}$$

to denote its "mean duration." If the mean duration of an activity is
defined to be the new duration, then our problem is reduced to a problem
of deterministic type and CPM may be used to obtain the total duration

of the project.

The so obtained project length M is certainly a possible duration of the project.   A natural question to ask is how to estimate the probability that the project is completed within M time units? Moreover, for any given N, how can one estimate the probability that the total duration of the project is less than or equal to N ?

In this chapter, we will introduce the so called PERT method for treating the above problem.   Underlying PERT are a number of statistical assumptions, and thus the reliability of these assumptions is certainly a question deserving study. However, PERT is a good topic for popularization because

1) it is easily accepted by workers and managers;

2) it is quite well suited for large projects which involve the participation of a multitude of people;

3) it is essentially a good scientific method for organization exercises involving several different industries.

§ 7.2 Estimation of the probability.

Hereafter we use t_o, t_m, t_p and t_e to denote respectively the optimistic duration, the most probable duration, the pessimistic duration and the mean duration of an activity, where t_e is defined by

$$t_e = \frac{t_o + 4 t_m + t_p}{6} .$$

The durations t_o-t_m-t_p are written on one side along the arc representing the activity and t_e is written on the other side, as in Figure 7.1.

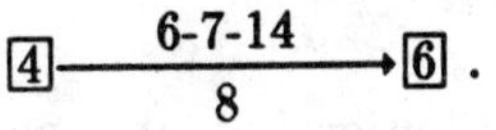

$$\boxed{4} \xrightarrow{\dfrac{6\text{-}7\text{-}14}{8}} \boxed{6} \ .$$

Figure 7.1

If t_e is used as the duration of the activity, then the problem of undeterministic type is reduced to a problem of deterministic type, and CPM can be used to find the total duration M of the project, and M is then called the mean duration of the project. Of course, the so obtained M is only a possible duration of the project. Now we proceed to introduce a method for estimating the probability p(N) that the project is completed within the amount of time N.

The variance of the duration of an activity is defined as

$$\tau_e \ = \ \left(\frac{t_p - t_o}{6} \right)^2$$

and the variance σ^2 of the project is defined as the sum of the τ_e values of the activities on the critical path. Then the probability

$$P(M + x\sigma)$$

is equal to

$$\Phi(x) \ = \ \frac{1}{\sqrt{2\pi}} \int_{-\infty}^{x} e^{-\frac{t^2}{2}} \, dt \ ,$$

where $\Phi(x)$ is given by Table 7.1. Since

$$\Phi(-x) \ = \ 1 - \Phi(x) \ ,$$

$\Phi(-x)$ can be derived from Table 7.1.

Table 7.1

Normal Distribution (distribution function)

x	$\Phi(x)$	x	$\Phi(x)$	x	$\Phi(x)$	x	$\Phi(x)$	x	$\Phi(x)$
0.00	0.500	0.60	0.726	1.20	0.885	1.80	0.964	2.40	0.992
0.05	0.520	0.65	0.742	1.25	0.894	1.85	0.968	2.45	0.993
0.10	0.540	0.70	0.758	1.30	0.903	1.90	0.971	2.50	0.994
0.15	0.560	0.75	0.773	1.35	0.911	1.95	0.974	2.55	0.995
0.20	0.579	0.80	0.788	1.40	0.919	2.00	0.977	2.60	0.995
0.25	0.599	0.85	0.802	1.45	0.926	2.05	0.980	2.65	0.996
0.30	0.618	0.90	0.816	1.50	0.933	2.10	0.982	2.70	0.997
0.35	0.637	0.95	0.829	1.55	0.939	2.15	0.984	2.75	0.997
0.40	0.655	1.00	0.841	1.60	0.945	2.20	0.986	2.80	0.997
0.45	0.674	1.05	0.853	1.65	0.951	2.25	0.988	2.85	0.998
0.50	0.691	1.10	0.864	1.70	0.955	2.30	0.989	2.95	0.998
0.55	0.709	1.15	0.875	1.75	0.960	2.35	0.991	3.00	0.999

Example. Consider the problem given by Figure 7.2.

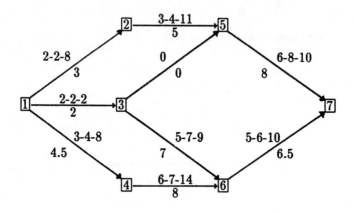

Figure 7.2

As a problem of deterministic type, we have the CP

$$\boxed{1} \xrightarrow{\;4.5\;} \boxed{4} \xrightarrow{\;8\;} \boxed{6} \xrightarrow{\;6.5\;} \boxed{7}$$

and its duration is $M = 19$. The mean and the variance of each activity on the CP are shown in Table 7.2.

Table 7.2

Activity	(1~4)	(4~6)	(6~7)	Total
Mean	4.5	8	6.5	19
Variance	$\left(\frac{8-3}{6}\right)^2 = \frac{25}{36}$	$\left(\frac{14-6}{6}\right)^2 = \frac{64}{36}$	$\left(\frac{10-5}{6}\right)^2 = \frac{25}{36}$	$\frac{114}{36}$

Hence the standard deviation of the duration of project is

$$\sqrt{\frac{114}{36}} \approx 1.8 .$$

From

$$19 = 19 + 1.8 \, x$$

we obtain

$$x = 0 ,$$

and so, from Table 7.1,

$$p(19) = \Phi(0) = 50 \% .$$

From

$$21 = 19 + 1.8 \, x ,$$

it follows that

$$x = \frac{2}{1.8} = 1.11 \cdots ,$$

and so, from Table 7.1,

$$p(21) = \Phi(1.11...) \approx 86 \% .$$

§ 7.3 Computation process.

Two tables are used to obtain the solution to the problem in section 7.1. The first table contains the following items.

1)   The names of the activities are put in the first column.

2)   The values for t_o, t_m, t_p and t_e are put in the second column.

3)   Use t_e as the duration of the activities. Use CPM to determine the early start time T_E and the late start time T_L of each activity, and put these quantities in the third column.

4)   The float of each activity is put in the fourth column. Connecting the arcs with zero float, we obtain the CP and the mean duration M of the project.

The second table contains the following items.

5)   The names of activities on the CP are put in the first column.

6)   The quantities t_o, t_p, $t_p - t_o$ and $(t_p - t_o)^2$ of the activities on the CP are put in the remaining columns. This way we obtain the standard deviation σ of the project duration.

For any given M , it follows from

$$M + x\sigma = N$$

that

$$x = \frac{N - M}{\sigma},$$

and so, using Table 7.1, we have that the probability P(N) that the project is finished within N days is equal to $\Phi(x)$.

Example. Consider the project shown in Figure 7.3.

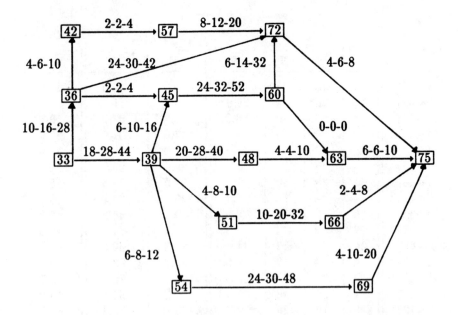

Figure 7.3

1) The first table is Table 7.3. From Table 7.3 we obtain immediately the CP

and the mean duration of the project is M = 95 .

2) The second table is Table 7.4. Hence the standard deviation of the project duration is

$$\sigma = \sqrt{\frac{2252}{36}} \approx \sqrt{62.5556} \approx 7.91 .$$

3) From

$$95 + 7.91\,x = 90 ,$$

it follows that

$$x = \frac{-5}{7.91} \approx -0.63 ,$$

Table 7.3

Activity	t_o	t_m	t_p	t_e	T_E	T_L	$T_L - T_E$
(33~36)	10	16	28	17.0	0	20.0	20.0
(33~39)	18	28	44	29.0	0	0	0
(36~42)	4	6	10	6.3	17.0	67.7	50.7
(36~45)	2	2	4	2.3	17.0	37.0	20.0
(36~72)	24	30	42	31.0	17.0	58.0	41.0
(39~45)	6	10	16	10.3	29.0	29.0	0
(39~48)	20	28	40	28.7	29.0	54.6	25.6
(39~51)	4	8	10	7.7	29.0	62.7	33.7
(39~54)	6	8	12	8.3	29.0	44.0	15.0
(42~57)	2	2	4	2.3	23.3	74.0	50.7
(45~60)	24	32	52	34.0	39.3	39.3	0
(48~63)	4	4	10	5.0	57.7	83.3	25.6
(51~66)	10	20	32	20.3	36.7	70.4	33.7
(54~69)	24	30	48	32.0	37.3	52.3	15.0
(57~72)	8	12	20	12.7	25.6	76.3	50.7
(60~63)	0	0	0	0	73.3	88.3	15.0
(60~72)	6	14	32	15.7	73.3	73.3	0
(63~75)	6	6	10	6.7	73.3	88.3	15.0
(66~75)	2	4	8	4.3	57.0	90.7	33.7
(69~75)	4	10	20	10.7	69.3	84.3	15.0
(72~75)	4	6	8	6.0	89.0	89.0	0
(75~75)	0	0	0	0	95.0	95.0	0

Table 7.4

Activity	t_o	t_p	$t_p - t_o$	$(t_p - t_o)^2$
(33~39)	18	44	26	676
(39~45)	6	16	10	100
(45~60)	24	52	28	784
(60~72)	6	32	26	676
(72~75)	4	8	4	16
			Total	2252

and thus, using Table 7.1, we have

$$p(90) \approx \Phi(-0.63) \approx 26\%.$$

4) From

$$95 + 7.91\,x = 100,$$

it follows that

$$x = \frac{5}{7.91} \approx 0.63,$$

and so, using Table 7.1, we have

$$p(100) \approx \Phi(0.63) \approx 74\%.$$

§ 7.4 An elementary approach.

A problem of undeterministic type is, as before, reduced to a problem of deterministic type and then CPM is used to find the CP of the latter problem. Suppose the CP consists of n activities $m_1, \ldots, m_n$. Let $t_o^{(i)}$, $t_m^{(i)}$ and $t_p^{(i)}$ denote the t_o, t_m and t_p of activity m_i. Now we proceed to evaluate the corresponding probability $p(N)$.

Let the weights of $t_o^{(i)}$, $t_m^{(i)}$ and $t_p^{(i)}$ be 1, 4 and 1. Consider the 3^n possible durations d for the CP

$$d = \sum_{i=1}^{n} d_i,$$

where d_i denotes one of $t_o^{(i)}$, $t_m^{(i)}$ and $t_p^{(i)}$. The weight of d, $\sigma(d)$ say, is defined to be the product of the weights of the d_i 's. Note that the sum of the weights of all 3^n possible d's is equal to 6^n. Now define the probability that the project is completed within N days as the ratio of the sum of the weights of all CP durations $\leq N$ and the sum of the weights of all CP durations, i.e.

$$p(N) = \frac{\sum\limits_{d \leq N} \sigma(d)}{6^n} \, .$$

Example. Consider the problem given in Figure 7.2. What is p(21)?

As a problem of deterministic type, the CP consists of the activities (1~4), (4~6) and (6~7), which are denoted by m_1, m_2 and m_3. The corresponding t_o, t_m and t_p are given in Table 7.5.

Table 7.5

Activity	t_o	t_m	t_p
m_1	3	4	8
m_2	6	7	14
m_3	5	6	10

Since 21 is larger than the mean project duration M, it is better to determine the probability q that the project will not be finished before the 22-th day, and then use

$$p(21) = 1 - q \, .$$

The CP durations d that are greater than or equal to 22 are the following.

$$t_o^{(1)} + t_p^{(2)} + t_o^{(3)} = 22, \qquad t_p^{(1)} + t_o^{(2)} + t_p^{(3)} = 24,$$

$$t_m^{(1)} + t_p^{(2)} + t_o^{(3)} = 23, \qquad\qquad t_p^{(1)} + t_m^{(2)} + t_p^{(3)} = 25,$$

$$t_p^{(1)} + t_p^{(2)} + t_o^{(3)} = 27, \qquad\qquad t_o^{(1)} + t_p^{(2)} + t_p^{(3)} = 27,$$

$$t_o^{(1)} + t_p^{(2)} + t_m^{(3)} = 23, \qquad\qquad t_m^{(1)} + t_p^{(2)} + t_p^{(3)} = 28,$$

$$t_m^{(1)} + t_p^{(2)} + t_m^{(3)} = 24, \qquad\qquad t_p^{(1)} + t_p^{(2)} + t_p^{(3)} = 32,$$

$$t_p^{(1)} + t_p^{(2)} + t_m^{(3)} = 28.$$

The sum of the corresponding weights 1, 4, 1, 4, 16, 4, 1, 4, 1, 4 and 1 is equal to 41 , and so

$$q = \tfrac{41}{216} \approx 19\ \%$$

and

$$p(21) \approx 1 - 19\ \% = 81\ \%.$$

§ 7.5 Remarks.

Underlying PERT are the statistical assumptions that the duration of each activity on the CP is a random variable following a Beta distribution over (t_o, t_p) with mean

$$t_e = \frac{t_o + 4\,t_m + t_p}{6}$$

and variance

$$c_e = \left(\frac{t_p - t_o}{6} \right)^2,$$

and that the total project duration is an approximately normally distributed random variable with mean M $(= \Sigma\, t_e)$ and standard deviation σ $(= \sqrt{\Sigma\, \tau_e}\,)$. Although the latter assumption may be regarded as the conclusion from the central limit theorem (A. M. Ljapunoff's theorem), the question how to check the assumptions inherent in the central limit theorem still remains open.

In the course of popularizing PERT, the meaning of the mean duration and the variance of an activity is explained as follows.

The probability of t_m is assumed to be twice that of t_o. Then

$$\frac{t_o + 2\, t_m}{3}$$

is regarded as the mean duration of the activity over (t_o, t_m). Similarly

$$\frac{2t_m + t_p}{3}$$

is used as the mean duration of the activity over (t_m, t_p). The mean duration and variance of an activity are taken to be

$$\frac{1}{2}\left(\frac{t_o + 2t_m}{3} + \frac{2t_m + t_o}{3} \right) = \frac{t_o + 4t_m + t_p}{6}$$

and

$$\frac{1}{2}\left(\left(\frac{t_o + 4t_m + t_p}{6} - \frac{t_o + 2t_m}{3} \right)^2 + \left(\frac{t_o + 4t_m + t_p}{6} - \frac{2t_m + t_p}{3} \right)^2 \right) =$$

$$= \left(\frac{t_p - t_o}{6} \right)^2.$$

The only assumption in the elementary approach is that the

weights of t_o, t_m and t_p are 1, 4 and 1, but the calculations for this method are comparatively long. More precisely, the number of elementary operations required is of exponential order.

References

Booz, Allen and Hamilton. Project PERT, Phase II, Special Projects Office, Bureau of Naval Weapons, Dept. of the Navy, Washington D.C., Nov. 1958.

Hua Loo Keng. Popular Lectures on Overall Planning Methods (with supplements), Chinese Industry Press, Beijing, 1965.

Hua Loo Keng, Chen De Quan, Ji Lei etc. (edited). A Collection of Achievements on Popularizing Optimum Seeking Methods in People's Republic of China, Vol. I, II, III, IV, Science and Tech. Document Press, Beijing, 1977.

Editor's note: Some additional references are the following.

Elmaghraby, S.E. Activity Networks: Project Planning and Control by Network Models, Wiley, 1977.

Moder, J.J., C.R. Phillips and E.W. Davis. Project Management with CPM, PERT and Precedence Diagramming, 3th ed., Van Nostrand, 1983.

Wiest, J.D., and F.K. Levy. A Management Guide to PERT/CPM, Prentice-Hall, 1977.

CHAPTER 8

MACHINE SCHEDULING

§ 8.1 Introduction.

Besides the problem of minimizing the total waiting time for one process, another problem connected with sequencing analysis that we have encountered in factories is the following machine scheduling problem:

<u>Problem</u>. Given n jobs $X_1, \ldots, X_n$ on s machines $A_1, \ldots, A_s$, where each job is to be processed successively on each machine according to the order $A_1 \ldots A_s$. The processing time, in terms of hours say, of each job on each machine is assumed to be given. How can the jobs be scheduled so that the time between the start of the first job and the completion of the last job is a minimum?

It was S.M. Johnson who solved the problem for the case $s = 2$, but for $s > 2$, we have only certain dirty methods (heuristic methods) with which merely more or less satisfactory solutions can be obtained. Therefore our discussion in this chapter is confined to the case $s = 2$.

§ 8.2 Two-machine problem.

Denote the two machines by A and B. Let a_i and b_i be the processing times of job X_i on A and on B, where a_i and b_i may take on the value zero. Johnson's algorithm may be stated as follows.

If the smallest number among $a_1, \ldots, a_n, b_1, \ldots, b_n$ appears

151

among the a's, say a_i, then the job X_i is scheduled as soon as possible. Otherwise, if it appears among the b's, say b_j, then X_j is scheduled as late as possible. In case of a tie, pick either job. Delete the scheduled job and repeat the above process.

Example. Suppose that there are five jobs $X_1, \ldots, X_5$ which will be processed on A and then on B. The processing times are given in Table 8.1.

Table 8.1: Processing times.

Machines

Jobs	A	B
X_1	1	5
X_2	8	3
X_3	3	9
X_4	4	2
X_5	7	5
Total	23	24

First of all, we will evaluate the total time needed for these five jobs if they are carried out in the order $X_1 X_2 X_3 X_4 X_5$. Let the start time of X_1 on A be $t = 0$. Then X_1 is finished at time $t = 1$ on machine A, and this is also the start time of X_1 on machine B. The end time of X_1 on B is $t = 1 + 5 = 6$. The start time of X_2 on A is $t = 1$ and its end time is $t = 1 + 8 = 9$. Therefore the start time and the end time of X_2 on B are $t = \text{MAX} \{ 6, 9 \} = 9$ and $t = 12$, etcetera. The results are shown in Table 8.2.

Twice the end time of the last job on B is called the total time of the schedule. The total sum of the processing times on the two machines is the machine time and the difference between total time and machine time is the idle time. So in this example, the total time, the machine time and the idle time for the schedule $X_1 X_2 X_3 X_4 X_5$ are $2 \times 28 = 56$, $23 + 24 = 47$ and 9 respectively.

Table 8.2

Machines

Jobs	A		B	
	In	Out	In	Out
X_1	0	1	1	6
X_2	1	9	9	12
X_3	9	12	12	21
X_4	12	16	21	23
X_5	16	23	23	28

Now we arrange a schedule according to Johnson's rule. The smallest number in Table 8.1 is 1, and thus job X_1 is scheduled first. Similarly, X_4 is scheduled last. The smallest number for the remaining jobs is now a tie of 3 between the job X_3 on machine A and the job X_2 on machine B. We arrange X_3 after X_1. Continuing in this manner we obtain the schedule $X_1 X_3 X_5 X_2 X_4$. The results are shown in Table 8.3. The total time for this schedule is 50 and the idle time is 3. This means that the schedule $X_1 X_3 X_5 X_2 X_4$ is better than the schedule $X_1 X_2 X_3 X_4 X_5$. We have the following result.

Theorem 8.1. A schedule with minimum total time can be obtained by Johnson's algorithm.

Table 8.3: A schedule using Johnson's algorithm.

Machines

Jobs	A In	A Out	B In	B Out
X_1	0	1	1	6
X_3	1	4	6	15
X_5	4	11	15	20
X_2	11	19	20	23
X_4	19	23	23	25

§ 8.3 A lemma.

Lemma 8.1. Suppose that three jobs, X_0, X_1, and X_2, are to be processed on machine A and then on machine B with processing times 0, a_1, a_2 and M, b_1, b_2. Assume that X_0 is arranged first. Then a schedule that arranges X_1 and X_2 according to Johnson's rule has a minimum total time.

Proof. Let the total time for the schedules $X_0 X_1 X_2$ and $X_0 X_2 X_1$ be $2f$ and $2g$. Then

$$f = \text{MAX} \{ \text{MAX} \{ M, a_1 \} + b_1, a_1 + a_2 \} + b_2$$

and

$$g = \text{MAX} \{ \text{MAX} \{ M, a_2 \} + b_2, a_1 + a_2 \} + b_1.$$

1) Suppose that $a_1 \le \text{MIN} \{ a_2, b_1, b_2 \}$. We have that

if $M < a_1$, then

$$f = \text{MAX} \{ a_1 + b_1, a_1 + a_2 \} + b_2, \quad g = a_2 + b_1 + b_2,$$

if $a_1 \leq M \leq a_2$, then

$$f = \text{MAX} \{ M + b_1, a_1 + a_2 \} + b_2, \quad g = a_2 + b_1 + b_2,$$

and if $M > a_2$, then

$$f = g = M + b_1 + b_2,$$

and therefore

$$f \leq g.$$

That is to say, the total time for the schedule $X_0\ X_1\ X_2$ does not exceed that for $X_0\ X_2\ X_1$. Similarly, the total time for the schedule $X_0\ X_2\ X_1$ is less than or equal to that for $X_0\ X_1\ X_2$ if $a_2 \leq \text{MIN} \{ a_1, b_1, b_2 \}$.

2) Suppose that $b_1 \leq \text{MIN} \{ a_1, a_2, b_2 \}$. We have that if $M < a_1$, then

$$f = a_1 + a_2 + b_2, \quad g = \text{MAX} \{ \text{MAX} \{ M, a_2 \} + b_2, a_1 + a_2 \} + b_1,$$

if $M < a_2$, then

$$f = a_1 + a_2 + b_2, \quad g = \text{MAX} \{ a_2 + b_2, a_1 + a_2 \} + b_1,$$

and if $M \geq \text{MAX} \{ a_1, a_2 \}$, then

$$f = \text{MAX}\ \{M + b_1, a_1 + a_2\} + b_2, \ g = \text{MAX}\ \{M + b_2, a_1 + a_2\} + b_1$$

and therefore

$$f \geq g.$$

That is to say, the total time for the schedule $X_0 X_2 X_1$ does not exceed that for $X_0 X_1 X_2$. Similarly, the total time for the schedule $X_0 X_1 X_2$ is less than or equal to that for $X_0 X_2 X_1$ if $b_2 \leq \text{MIN}\ \{ a_1, a_2, b_1 \}$. The lemma is proved. □

§ 8.4 Proof of Theorem 8.1.

For the case $n = 2$, add an idle job X_0 to be scheduled first, with the processing times $0, M$. Then the theorem follows immediately from Lemma 8.1 if we take $M = 0$. Suppose that $n \geq 3$ and that the theorem holds if the number of jobs does not exceed $n-1$. It suffices to prove that for any schedule σ of n given jobs $X_1, \ldots, X_n$, there exists a schedule obtained by Johnson's rule whose total time does not exceed the total time for σ.

Without loss of generality, we may suppose that the order of σ is $X_1, \ldots, X_n$. First of all, we use Johnson's rule to schedule the first $n-1$ jobs $X_1, \ldots, X_{n-1}$, and obtain

$$X_{i_1} \ \ldots \ X_{i_{n-1}}.$$

It follows by induction that the total time for $X_{i_1} \ \ldots \ X_{i_{n-1}} X_n$ does not exceed the total time for σ.

Let $2N$ be the total time of the schedule $X_{i_1} \ \ldots \ X_{i_{n-2}}$. Let

$$M = N - a_{i_1} - \cdots - a_{i_{n-2}}.$$

Furthermore let the processing times of an idle job X_0 be 0, M. Schedule the jobs $X_{i_{n-1}}$ and X_n according to Johnson's rule, and we obtain

$$X_{j_{n-1}} X_{j_n}.$$

Then it follows from Lemma 8.1 that the total time for $X_0 \, X_{j_{n-1}} X_{j_n}$ does not exceed that of $X_0 \, X_{i_{n-1}} X_n$. That is to say, the total time for the schedule $X_{i_1} \, \ldots \, X_{i_{n-2}} X_{j_{n-1}} X_{j_n}$ is less than or equal to the total time for the schedule $X_{i_1} \, \ldots \, X_{i_{n-1}} X_n$, and consequently, to the total time for σ. Scheduling the jobs $X_{i_1}, \ldots, X_{i_{n-2}}, X_{j_{n-1}}$ according to Johnson's rule, we obtain

$$X_{k_1} \, \ldots \, X_{k_{n-1}}.$$

It follows by induction that the total time for $X_{k_1} \, \ldots \, X_{k_{n-1}} X_{j_n}$ does not exceed that for $X_{i_1} \, \ldots \, X_{i_{n-2}} X_{j_{n-1}} X_{j_n}$, and therefore that for σ. Since the schedule $X_{k_1} \, \ldots \, X_{k_{n-1}} X_{j_n}$ is in agreement with Johnson's rule, the theorem follows. $\Box$

Remark. The n jobs are always processed continuously on A but usually there are many floats on B. Let the total time for a schedule be $2N$. Let

$$M = N - \sum_{i=1}^{n} b_i.$$

Then the jobs can also be processed continuously on B if the start time of the first job on B is defined to be $t = M$. Hence the machine B can be used for other work before time $t = M$.

References.

Johnson, S.M. Optimal Two and Three Stage Production Schedules with Setup Time Included, <u>Naval</u> <u>Reasearch</u> <u>Logistics</u> <u>Quarterly</u>, 1, 1954, 61-68.

Wu Fang. <u>Remarks</u> <u>on</u> <u>Johnson's</u> <u>rule</u> <u>for</u> <u>the</u> <u>Two-Machine</u> <u>Problem</u>, 1981, unpublished.

Yu Ming-I and Han Ji Ye, Some Mathematical Problems in Sequencing Analysis, <u>Math.</u> <u>Recognition</u> <u>and</u> <u>Practice</u>, 3, 1976, 59-70; 4, 1976, 62-77. (See the remarks concerning this journal at the end of Chapter 2).

Editor's note: The scheduling problem in this chapter is a so called flow shop system and, as shown in this chapter, can be solved rapidly. For many scheduling problems no polynomial time algorithm to determine an optimal schedule is known, these problems are NP-hard. There exist a vast amount of literature on scheduling problems, much of it dealing with computational complexity issues and heuristic algorithms. Additional references are given below, where the second reference is survey article containing an extensive bibliography.

Bellman, R., A.O. Esogbue and I.Nabeshima. <u>Mathematical</u> <u>Aspects</u> <u>of</u> <u>Scheduling</u> <u>and</u> <u>Applications</u>. Pergamon Press, 1982.

Błażewicz, J. Selected Topics in Scheduling Theory. 1-60. In Martello, S., G. Laporte, M. Minoux and C. Ribeiro, <u>Surveys</u> <u>in</u> <u>Combinatorial</u> <u>Optimization</u>, Annals of Discrete Mathematics, 31, North Holland, 1987.

CHAPTER 9

THE TRANSPORTATION PROBLEM (GRAPHICAL METHOD)

§ 9.1 Introduction.

In 1958 we started popularizing mathematical methods to the industrial departments in China. The first problem they asked us about is the following transportation problem. Suppose that it is desired to devise a transportation schedule for the distribution of a homogeneous product, for example distributing wheat by railway. Suppose that there are m sources $A_1, \cdots, A_m$ with the amounts of supply $a_1, \cdots, a_m$ (in terms of tons, say) and n destinations (consumer places) $B_1, \cdots, B_n$ with the amounts of demand $b_1, \cdots, b_n$. The total supply at all sources is assumed to be equal to the total demand at all destinations, i.e.

$$\sum_{i=1}^{m} a_i = \sum_{j=1}^{n} b_j .$$

Suppose that the distance between A_i and B_j is c_{ij} (in kilometres, say) and the amount being shipped from A_i to B_j is x_{ij}. Then the total transportation cost (in terms of ton-kilometres) as a function of the vector $\vec{x}$ of shipping amounts is given by

$$f(\vec{x}) = \sum_{i=1}^{m} \sum_{j=1}^{n} c_{ij} x_{ij} .$$

The a_i's, b_j's, c_{ij}'s and x_{ij}'s are assumed to be integral. The problem

159

is how to choose x_{ij} $(1 \le i \le m, 1 \le j \le n)$ such that $f(\vec{x})$ is a minimum.

Remark. If c_{ij} denotes the transportation cost in Yuan (the Chinese currency) per ton from A_i to B_j, then $f(\vec{x})$ denotes the total transportation cost in Yuan.

This is the well-known transportation problem of Hitchcock and Kantorowitch in linear programming, and it can be solved by the simplex method which will be discussed in the next chapter. The workers who set up grain transportation schedules for railways in the Chinese transportation department proposed the above problem independently, based on their experiences of many years, and they also gave a method to solve the problem with the aid of a map. We will now illustrate their method by an example.

Suppose that there are four sources A_1, A_2, A_3, A_4 with supplies 20, 20, 60, 100 tons and five destinations B_1, B_2, B_3, B_4, B_5 with demands 50, 30, 30, 70, 20 tons. The geographical locations of the sources and destinations are depicted in Figure 9.1.

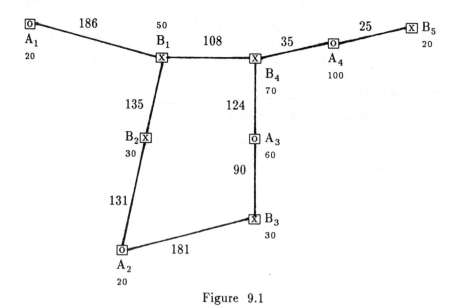

Figure 9.1

Sources are denoted by the symbol ⊡ and destinations by the symbol ⊠. The number next to a source or a destination represents the supply or demand. The distances between neighboring cities and towns are given along the connecting lines.

Table 9.1: A transportation schedule.

Destinations

		B_1	B_2	B_3	B_4	B_5	Supply
	A_1		20				20
	A_2	20					20
Sources	A_3		10	30		20	60
	A_4	30			70		100
	Demand	50	30	30	70	20	

For any given transportation schedule such as Table 9.1, one may draw a map to represent it. Suppose that m tons of a certain good are transported from A to B. A flow vector is drawn alongside the line segment representing the railway from A to B with the amount of flow m written above the flow vector, as in Figure 9.2.

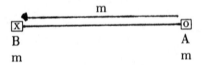

Figure 9.2

If there is more than one flow vector in the same direction along a line segment, a single flow vector is drawn indicating the total amount of transported goods, as shown in Figure 9.3.

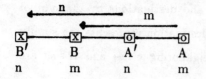 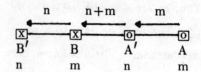

Figure 9.3

When this is done for all line segments, a complete map corresponding with the schedule is obtained and is called <u>the</u> <u>flow</u> <u>diagram</u> of the schedule. For example, Figure 9.4 is a flow diagram of the schedule given in Table 9.1. It is evident that the transportation costs $f(\vec{x})$ for schedules with the same flow diagram are equal.

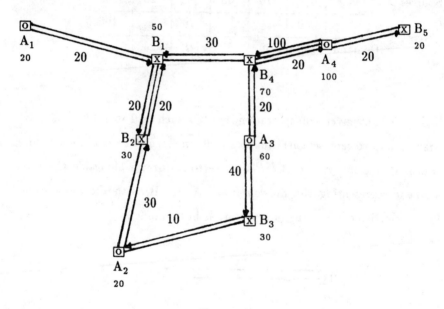

Figure 9.4

Two flow vectors that appear in opposite directions on a segment are called <u>counter</u> <u>flow</u> <u>vectors</u>. For example, the segments $B_1 B_2$ and $A_4 B_4$ in Figure 9.4 have counter flow vectors. If a flow diagram has counter flows, the corresponding schedule is certainly not the best. In such a case the schedule may be changed to obtain a better one that contains no counter flows. For example, Figure 9.4 can be changed into Figure 9.5, and the corresponding transportation schedule is given in Table 9.2.

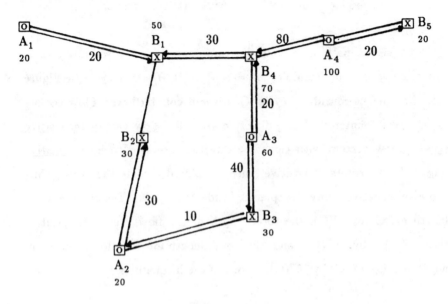

Figure 9.5

Table 9.2: A transportation schedule without counter flows.

Destinations

		B_1	B_2	B_3	B_4	B_5	Supply
	A_1	20					20
	A_2		20				20
Sources	A_3		10	30	20		60
	A_4	30			50	20	100
	Demand	50	30	30	70	20	

The difference between the transportation costs of the schedules given in

Tables 9.1 and 9.2 is equal to the cost wasted on the segments with counter-flows in Figure 9.4, i.e.

$$2 \, (135 \times 20 + 35 \times 20) = 6800 \quad \text{(ton-kilometers)}.$$

A closed path in a flow diagram that does not self-intersect is called a cycle. For example, $B_1B_2A_2B_3A_3B_4B_1$ is a cycle in Figure 9.1. Now let us consider a cycle C without counterflows. Flow vectors of C with counter-clockwise direction are called flow vectors in positive sense. Flow vectors with clockwise direction are said to have negative sense. Flow vectors in positive sense are placed outside the cycle, while those in negative sense are placed inside the cycle. Denote the total length of C by $\ell(C)$, the total length of the flow vectors in positive sense of C by $\ell^+(C)$ and the total length of the flow vectors in negative sense of C by $\ell^-(C)$. If one of the inequalities

$$\ell^+(C) \leq \tfrac{1}{2}\ell(C) \ \text{ and } \ \ell^-(C) \leq \tfrac{1}{2}\ell(C)$$

is not satisfied, the cycle is called non-normal. One can prove that if a non-normal cycle appears in a flow diagram, then the corresponding transportation schedule is certainly not optimal. In such a case the schedule should be changed so as to obtain a better one, one for which the flow diagram contains no non-normal cycle. For example, the cycle $B_1B_2A_2B_3A_3B_4B_1$ in the flow diagram of Figure 9.5 has length 769 kilometres, and the total length of the flow vectors in negative sense is 402 kilometres, which is greater than one-half of 769 kilometres. Therefore it is a non-normal cycle. Now the flow diagram should be changed to a diagram in which the non-normal cycle has disappeared. For example, Figure 9.5 can be changed to Figure 9.6 by prolonging the flow in positive sense and shortening the flow in negative sense.

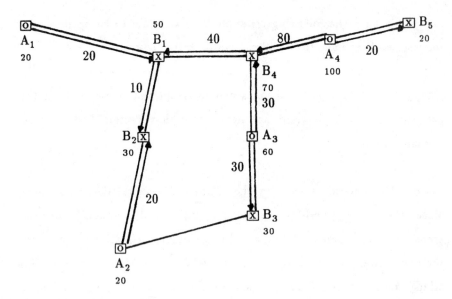

Figure 9.6

The transportation schedule corresponding to Figure 9.6 is show in Table 9.3, and the difference between the transportation costs of the schedules given in Tables 9.2 and 9.3 is equal to the cost wasted on the non-normal cycle of Figure 9.5, i.e.

$$10\,(131 + 181 + 90) - 10\,(124 + 108 + 135) = 350 \quad \text{(ton-kilometres)}.$$

Table 9.3: A transportation schedule without non-normal cycles.

Destinations

		B_1	B_2	B_3	B_4	B_5	Supply
	A_1	10	10				20
	A_2		20				20
Sources	A_3			30	30		60
	A_4	40			40	20	100
	Demand	50	30	30	70	20	

We have the following criterion for a schedule to be optimal, i.e. for a schedule to have the smallest transportation cost.

Theorem 9.1. A transportation schedule is optimal if and only if the corresponding flow diagram has no counter flow vectors and no non-normal cycles.

This method is used in the Chinese transportation department since the 1950's and is called the "graphical method" for the transportation problem. They asked us to provide a mathematical proof for their criterion, and this was done by Wan Zhe Xian as well as Gui Xiang Yun and Xu Guo Zhi (see Wan Zhe Xian). This method is obviously inconvenient if the number of sources, destinations or cycles in the flow diagram is comparatively large. In this case the well-known simplex method has more advantages, in particular it is easy to write a computer program for this method. So we also introduced them to the simplex method in 1958.

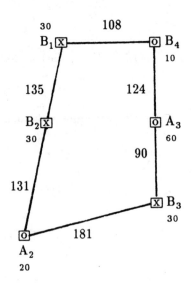

Figure 9.7

Remarks. 1) The branching lines of a map can be omitted in the following way. The various supplies and demands on a branching line may be concentrated in the intersection point of the branching line and the cycle. For example, in Figure 9.1 the line $B_4 A_4 B_5$ can be omitted by changing B_4 into a source with a net supply of 10 tons, and the line $A_1 B_1$ can also be neglected by considering B_1 to be a destination with a net demand of 30 tons. In this manner we obtain a map containing no branching lines (see Figure 9.7).

2) Guan Mei Gu and Li Zhi Jie proved that the number of cycles that should be examined in applications of Theorem 9.1 can be decreased in some cases.

§ 9.2 One cycle.

It is evident that a flow diagram for an optimal transportation schedule contains no counter flow vectors, and therefore we consider only flow diagrams without counter flows. Suppose furthermore that the flow diagram contains only cycles, i.e. there are no branching lines. If such a flow diagram contains no non-normal cycle, then it is called a <u>normal flow diagram</u>, otherwise it is a non-normal flow diagram.

In this section, we consider only the case where the flow diagram consists of a single cycle. For simplicity, we assume that the demand of each destination is 1 unit. In fact, if a destination demands m units, then we may instead imagine a series of m destinations, each with a demand of 1 unit, situated at the original destination with distance 0 between any two such destinations.

Proof of Theorem 9.1 (case of one cycle).

1) Suppose we have a transportation schedule with a non-normal

flow diagram C. Without loss of generality, assume that

$$\ell^+(C) > \tfrac{1}{2}\ell(C) . \tag{9.1}$$

We may draw a new flow diagram C' which differs from C in the sense that the end point of each maximal path of flow vectors in positive sense in C will now be supplied by a flow vector in negative sense in C', while the other destinations will be supplied by the same source in C' as in C. That is, the amount of flow along line segments whose flow vector has positive sense in C will be reduced by 1, while the amount of flow along line segments whose flow vector has negative sense in C will be increased by 1, and line segments with zero flow in C will now have flow 1 in negative sense. This process of changing from C to C' is called the process of shrinking the flow vectors in positive sense by one station or, simply, substitution of the flow diagram. The marginal transportation cost for transporting one unit to each end point of the maximal paths of flow vectors in positive sense of C is equal to $\ell^+(C)$ in C, while the marginal cost of supplying the same destinations in C' is equal to $\ell(C) - \ell^+(C)$. Using (9.1), we have

$$\ell^+(C) > \tfrac{1}{2}\ell(C) > \ell(C) - \ell^+(C) .$$

Thus the transportation schedules corresponding with flow diagram C' are better than those corresponding with C. This proves that a schedule with a non-normal diagram is never optimal. Since there are only finitely many relevant flow diagrams (there is no need to consider flow diagrams for which all line segments have flow vectors in positive sense), a normal flow diagram always exists.

    2)   Next, we will prove that all transportation schedules with normal flow diagrams are optimal, i.e. they have the same transportation

cost. Note that a flow diagram is completely determined once the amount of flow on a particular line segment is known. Let us arrange the flow diagrams in the order

$$C_1, \cdots, C_i, C_{i+1}, \cdots, C_n,$$

where C_1 is a flow diagram without flow vectors in negative sense, C_n is a flow diagram without flow vectors in positive sense, and each C_{i+1} is obtained from the preceding C_i by the process of shrinking the flow vectors in positive sense of C_i by one station, where $1 \leq i \leq n-1$. Note that all normal flow diagrams are included in this list of flow diagrams. Let C_s be the normal flow diagram with smallest subscript and C_r the normal flow diagram with greatest subscript. Then the normal flow diagrams are precisely

$$C_s, C_{s+1}, \cdots, C_t .$$

If $s = t$, then there is only one normal flow diagram and therefore the theorem holds. Now suppose that $s < t$. It suffices to show that any two neighboring flow diagrams C_k and C_{k+1} $(s \leq k < t)$ have the same transportation cost. The line segments whose flow vector have negative sense in C_{k+1} are precisely the line segments whose flow vector did not have positive sense in C_k, thus

$$\ell^-(C_{k+1}) = \ell(C) - \ell^+(C_k) .$$

Since

$$\ell^+(C_k) \leq \tfrac{1}{2}\ell(C)$$

and

$$\ell^-(C_{k+1}) \leq \tfrac{1}{2}\ell(C) ,$$

we have

$$\ell^+(C_k) = \ell^-(C_{k+1}) = \tfrac{1}{2}\ell(C).$$

This proves that C_k and C_{k+1} have the same transportation cost, and the theorem is proved. □

§ 9.3 Proof of Theorem 9.1.

Similar to the case of one cycle, it suffices to show that all schedules with normal flow diagram have the same transportation cost.

Let F and F' be two distinct normal flow diagrams, which, of course, correspond to different schedules. When comparing F with F' we may suppose that F and F' have no common flow vectors. Otherwise, we may reduce in both flow diagrams the flow on line segments that have common flow vectors by the smaller of the two flow vectors. We obtain flow diagrams G and G' without common flow vectors. If the transportation costs for G and G' are equal, then they are also equal for F and F'. In order to ensure that G and G' represent flow diagrams corresponding with some transportation problem, we must adjust the supplies and demands at the endpoints of the line segments with common flow vectors in F and F'. If the flow vector on a line segment is reduced by δ then the supply (resp. demand) at the starting point must be reduced (resp. increased) if the starting point was a source (resp. a destination), and a similar adjustment must be made at the endpoint. Note that an endpoint that was previously a source (resp. a destination) may now be a destination (resp. a source) in the new transportation problem. If a line segment has a flow vector in G then it has a flow vector in the same direction in F, and thus G (and G')

will be a normal flow diagram.

Let N denote the total amount of supply at all sources of F. If $N = 0$, the theorem holds clearly. Suppose that $N > 0$ and that the theorem is true for all non-negative integers less than N. Now we proceed to prove that the theorem holds for N also. Let us start with a source A_1. Suppose that, in F, at least one ton of goods is transported from A_1 to destination B_1. Then, in F$'$, at least one ton of the goods needed by B_1 is supplied by some source A_2 (A_2 could possibly be the same as A_1, but then an alternative route is used). Then in F, A_2 supplies some destination B_2. Continuing in this way, we obtain a sequence

$$A_1 \ B_1 \ A_2 \ \cdots .$$

Consider the first location that is repeated in this sequence. Without loss of generality we assume that it is a destination, B_i say. We obtain a closed path Z between the first and second occurrence of B_i,

$$Z: \ B_i \ A_{i+1} \ \cdots \ B_{j-1} \ A_j \ B_i .$$

Along Z, the various deliveries of goods according to F go around in one sense, while all deliveries of goods according to F$'$ go around the other way. We obtain two new flow diagrams F_1 and F_1' by reducing the supplies at $A_{i+1}, \ldots, A_j$ by one ton, reducing the demands at B_i $\cdots, B_{j-1}$ by one ton, reducing the flow vectors of F on the $A_s B_s$ sections of Z (including $A_j B_i$) by one ton and reducing the flow vectors of F$'$ on the $B_s A_{s+1}$ sections of Z by one ton. Clearly, F_1 and F_1' are normal flow diagrams without common flow vectors and the total supply at all sources is smaller than for F and F$'$. By the induction hypothesis, F_1 and F_1' have the same transportation cost.

Z may be decomposed into several cycles, $C_1, \cdots, C_m$ such that in each cycle C_k the flow vectors of F have the same sense and the flow vectors of F' all have the opposite sense. Since both F and F' are normal, the transportation costs for all one ton deliveries according to F and according to F' in C_k are the same, i.e. both are equal to $\frac{1}{2}\ell(C_k)$ (see § 9.2). Hence the total transportation cost for F (respectively F') equals the sum of the total transportation cost of F_1 (respectively F_1') and $\frac{1}{2}\ell(Z)$. Thus F and F' have the same transportation cost. The theorem is proved. □

References

Wan Zhe Xian. A proof for a Graphic Method for solving Transportation Problem, Scientia Sinica, 7, 1962, 889-894.

Wan Zhe Xian and Wang Yuan. Mathematical Methods in Transportation Problem, Science Press, Beijing, 1959.

Yu Ming-I, Wan Zhe Xian, Wang Yuan etc. (edited). The Theory and Application of Linear Programming, People's Education Press, Beijing, 1959.

Editor's note: Additional references on related graphical methods for the transportation problem, the transshipment problem and the minimum cost network flow problem are the following.

Ford, L.R. Jr. , and D.R. Fulkerson. Flows in Networks, Princeton University Press, 1962.

Lawler, E.L. Combinatorial Optimization: Networks and Matroids, Holt, Rinehart and Winston, 1976.

CHAPTER 10

THE TRANSPORTATION PROBLEM (SIMPLEX METHOD)

§ 10.1 Introduction.

As far as we know, the graphical method is inefficient if the number of sources, destinations or cycles in a map is comparatively large, and so, in 1958, the simplex method (Dantzig's method) was also introduced and popularized to the workers in the transportation departments of China. In this chapter, we will illustrate the method following the lecture notes written by Yu Ming-I, Wan Zhe Xian and Wang Yuan in 1958. The notations of the preceding chapter are also used here. Now let us start with the following example.

1) Set up an initial transportation schedule (using the smallest element method). First we will make a table (Table 10.1 for our example) that includes the supplies and demands at the sources and destinations and the per unit transportation costs c_{ij}, e.g. in terms of Yuan per ton.

Table 10.1: Transportation costs, supplies and demands.

Destinations

		B_1	B_2	B_3	B_4	Supply
	A_1	3	11	3	10	7
Sources	A_2	1	9	2	8	4
	A_3	7	4	10	5	9
	Demand	3	6	5	6	

Next, we search the table for the smallest per unit transportation cost c_{ij} and make a maximum allocation from A_i to B_j, and then repeat this procedure until all the units are shipped. Since the smallest number in Table 10.1 is 1, which is located at position $(2,1)$, B_1 is supplied by A_2 by as many units as possible, and therefore the position $(2,1)$ in a second table (a table that will contain an initial transportation schedule) will be filled by the number 3. Since the demand of B_1 is fullfilled, the positions $(1,1)$ and $(3,1)$ should be left blank as shown in Table 10.2.

Table 10.2

Destinations

	B_1	B_2	B_3	B_4
A_1				
A_2	3			
A_3				

Sources

The smallest number in the remaining positions of Table 10.1 is 2, located at position $(2,3)$, and thus B_3 is supplied with as many units as possible from A_2. The position $(2,3)$ in Table 10.2 is filled with 1 and the positions $(2,2)$ and $(2,4)$ are left blank since the supply at A_2 is exhausted. We obtain Table 10.3.

Table 10.3

Destinations

	B_1	B_2	B_3	B_4
A_1				
A_2	3		1	
A_3				

Sources

Continuing this process, we obtain the initial transportation schedule given in Table 10.4.

Table 10.4: Initial transportation schedule.

Destinations

		B_1	B_2	B_3	B_4	Supply
	A_1			4	3	7
Sources	A_2	3		1		4
	A_3		6		3	9
	Demand	3	6	5	6	

Remark. In general, after each allocation from a source to a destination, one source or one destination is eliminated. Since the total supply is assumed to be equal to the total demand, the initial transportation schedule is obtained after $n+m-1$ steps. Notice that the amount of allocation may be equal to zero.

2) Optimality criterion for transportation schedules (the criterion number method). Create a new table containing the transportation costs corresponding with the positions of the allocations in the initial transportation schedule, namely the positions (1,3), (1,4), (2,1), (2,3), (3,2) and (3,4) (see Table 10.5). Then a criterion number is chosen for each row and each column such that each entry in Table 10.5 is the sum of the two criterion numbers corresponding with its row and its column. For example, putting 0 below B_1 implies that 1 must be placed to the right of A_2, and consequently, 1 goes below B_3, 2 to the right of A_1, 8 below B_4, -3 to the right of A_3 and 7 below B_2. Each blank location in Table 10.5 is filled with the sum of the criterion numbers of the corresponding row and column, and we have Table 10.6.

Table 10.5: Criterion numbers.

Destinations

	B_1	B_2	B_3	B_4	Criterion numbers
A_1			3	10	2
A_2	1		2		1
A_3		4		5	−3
Criterion numbers	0	7	1	8	

Sources (label at left of A_1–A_3 rows)

Table 10.6

Destinations

	B_1	B_2	B_3	B_4	Criterion numbers
A_1	2	9	3	10	2
A_2	1	8	2	9	1
A_3	−3	4	−2	5	−3
Criterion numbers	0	7	1	8	

Sources

Subtracting the entries in Table 10.6 from the corresponding transportation costs c_{ij}, we obtain Table 10.7.

Table 10.7: Criterion table.

Destinations

	B_1	B_2	B_3	B_4
A_1	1	2	0	0
A_2	0	1	0	−1
A_3	10	0	12	0

Sources

If every number in the criterion table is non-negative, then the corresponding transportation schedule is optimal. Otherwise, the schedule can be changed to one that is at least as good.

3) <u>Substitution</u> of the transportation schedule. Since the entry in location (2,4) of Table 10.7 is negative, the corresponding transportation schedule given in Table 10.4 should be changed to a better one. We draw, in Table 10.4, a closed path Z through (2,4) whose edges are all parallel to the "coordinate axis". Except for (2,4), its vertices are located at positions that contain an entry, i.e.

$$Z: \quad (2,4) \quad (1,4) \quad (1,3) \quad (2,3) \quad (2,4),$$

as shown in Table 10.8. Starting from (2,4), the remaining vertices of Z are labeled 1, 2, 3 and so on. Thus the odd vertices are (1,4) and (2,3), while the others are the even vertices. Choose the smallest number among the entries at the odd vertices of Z, i.e. 1 in the position (2,3).

Table 10.8

Destinations

		B_1	B_2	B_3	B_4	Supply
	A_1			4	3	7
Sources	A_2	3		1		4
	A_3		6		3	9
	Demand	3	6	5	6	

The new transportation schedule will have a blank at location (2,3). The entries in the odd vertices are reduced by 1, while the entries in the even vertices are increased by 1. We obtain a new transportation schedule, shown in Table 10.9.

178

Table 10.9: New transportation schedule.

Destinations

	B_1	B_2	B_3	B_4	Supply
A_1			5	2	7
A_2	3			1	4
A_3		6		3	9
Demand	3	6	5	6	

Sources (label at left of A_1–A_3 rows)

The transportation cost saved by this schedule compared with the original schedule is equal to

$$(4{\times}3 + 1{\times}2 + 3{\times}10) - (5{\times}3 + 2{\times}10 + 1{\times}8) = 1 \ \text{(Yuan)},$$

and the criterion table corresponding with Table 10.9 is given in Table 10.10.

Table 10.10: Criterion table.

Destinations

	B_1	B_2	B_3	B_4
A_1	0	2	0	0
A_2	0	2	1	0
A_3	9	0	12	0

Sources (label at left of A_1–A_3 rows)

Since every entry in Table 10.10 is non-negative, the schedule given in Table 10.9 is optimal.

Remarks. 1) Any transportation schedule with $m+n-1$ allocations and whose corresponding flow diagram contains no counter flows may be regarded as an initial transportation schedule for the simplex method. In practice it is sufficient to take a few significant digits for the c_{ij}'s.

2) Tian Ke Jun and Xu Zhong Lin wrote a computer program for the above procedure.

In the following sections, we will provide a rigorous proof for the above method.

§ 10.2 Eliminated unknowns and feasible solutions.

Let c_{ij} $(1 \le i \le m, 1 \le j \le n)$, a_i $(1 \le i \le m)$ and b_j $(1 \le j \le n)$ be positive numbers satisfying

$$\sum_{i=1}^{m} a_i = \sum_{j=1}^{n} b_j . \tag{10.1}$$

Problem. Determine a set of numbers x_{ij}, $1 \le i \le m$, $1 \le j \le n$, for which the objective function

$$F(\vec{x}) = \sum_{i=1}^{m} \sum_{j=1}^{n} c_{ij} x_{ij} \tag{10.2}$$

attains its minimum, and where the x_{ij}'s satisfy the constraints

$$\sum_{j=1}^{n} x_{ij} = a_i , \quad 1 \le i \le m,$$

$$\tag{10.3}$$

$$\sum_{i=1}^{m} x_{ij} = b_j , \quad 1 \le j \le n$$

and

$$x_{ij} \ge 0, \quad 1 \le i \le m, \ 1 \le j \le n. \tag{10.4}$$

There are mn unknowns in (10.3), and it is easily proved that any $m+n-1$ equations are linearly independent. Thus in (10.3) there are $m+n-1$ unknowns, $y_1, \ldots, y_{m+n-1}$ say, that can be represented uniquely in terms of the remaining unknowns, $x_1, \ldots, x_{mn-m-n+1}$ say, as follows:

$$y_i = k_i + \sum_{j=1}^{mn-m-n+1} \epsilon_{ij} x_j, \quad 1 \leq i \leq m+n-1. \quad (10.5)$$

The unknowns $y_1, \ldots, y_{m+n-1}$ are called <u>the eliminated unknowns</u> and $x_1, \ldots, x_{mn-m-n+1}$ are <u>the remaining unknowns</u>. If

$$k_i \geq 0, \quad 1 \leq i \leq m+n-1,$$

then $\{y_1, \ldots, y_{m+n-1}\}$ is called <u>a set of positive eliminated unknowns</u>, and therefore

$$y_i = k_i, \quad 1 \leq i \leq m+n-1,$$

$$x_j = 0, \quad 1 \leq j \leq mn-m-n+1.$$

is a solution satisfying (10.3) and (10.4), and is called <u>a feasible solution</u>. A feasible solution $\vec{x}$ is called <u>an optimal solution</u> if F attains its minimum at $\vec{x}$.

In fact, the smallest element method can be used to determine a set of positive eliminated unknowns. Consider the $(m+1) \times (n+1)$ matrix

$$D = \begin{bmatrix} C & A \\ B & O \end{bmatrix},$$

where

$$C = \left[c_{ij} \right]_{1 \le i \le m,\ 1 \le j \le n},$$

$$A = \left[a_1 \ \cdots \ a_m \right]^T \ \text{and} \ B = \left[b_1 \ \cdots \ b_n \right].$$

Let $c_{i_0 j_0} = \underset{C}{\text{MIN}}\ c_{ij}$. Without loss of generality assume that $a_{i_0} \le b_{j_0}$. Then

$$x_{i_0 j_0} = a_{i_0} - x_{i_0 1} - \cdots - x_{i_0, j_0 - 1} - x_{i_0, j_0 + 1} - \cdots - x_{i_0 n}.$$

Next consider the $m \times (n+1)$ matrix

$$D_1 = \begin{bmatrix} C_1 & A_1 \\ B_1 & O \end{bmatrix},$$

where C_1 and A_1 are obtained by omitting the i_0-th row of C and of A respectively, and B_1 is obtained by replacing the j_0-th entry b_{j_0} of B by $b_{j_0} - a_{i_0}$. Let $c_{i_1 j_1} = \underset{C_1}{\text{MIN}}\ c_{ij}$. Suppose that $b_{j_1} < a_{i_1}$. Then

$$x_{i_1 j_1} = b_{j_1} - x_{1 j_1} - \cdots - x_{i_1 - 1, j_1} - x_{i_1 + 1, j_1} - \cdots - x_{m j_1}$$

if $j_1 \ne j_0$, and

$$x_{i_1 j_1} = b_{j_0} - a_{i_0} - x_{1 j_0} - \cdots - x_{i_0 - 1, j_0} - x_{i_0 + 1, j_0} - \cdots$$

$$- x_{i_1 - 1, j_0} - x_{i_1 + 1, j_0} - \cdots - x_{m j_0} +$$

$$+ x_{i_0 1} + \cdots + x_{i_0, j_0 - 1} + x_{i_0, j_0 + 1} + \cdots + x_{i_0 n}$$

if $j_1 = j_0$. And then consider the $m \times n$ matrix

$$D_2 = \begin{bmatrix} C_2 & A_2 \\ B_2 & O \end{bmatrix},$$

where C_2 and B_2 are obtained by deleting the j_1-th column of C_1 and B_1 respectively, and A_2 is obtained by replacing a_{i_1} of A_1 by $a_{i_1} - b_{j_1}$. Continuing this process and in view of (10.1), we obtain a set of $m+n-1$ unknowns that are expressed in the remaining unknowns and for which the constant terms are non-negative, i.e. we have a set of positive eliminated unknowns.

§ 10.3 Criterion numbers.

Let $X = \left[x_{ij} \right]_{1 \leq i \leq m,\ 1 \leq j \leq n}$. A closed broken line in X whose edges are parallel to the "coordinate axes" (alternating between horizontal and vertical), whose vertices are certain x_{ij}'s, and such that each row and column of X contains at most one edge, is called a closed path of X. For example, $x_{21}x_{23}x_{13}\ x_{12}x_{32}\ x_{31}x_{21}$ is a closed path.

Theorem 10.1. A set Y with $m+n-1$ elements y_i $(1 \leq i \leq m+n-1)$ of X forms a set of eliminated unknowns if and only if there does not exist a closed path of X whose vertices belong to Y.

Proof. 1) Suppose that $Y = \{ y_i : 1 \leq i \leq m+n-1 \}$ is a set of eliminated unknowns and that there exists a closed path of X whose vertices are in Y, say

$$y_1 y_2 \cdots y_{2r} y_1.$$

Suppose the expressions for these unknowns in terms of the remaining unknowns are

$$y_i = k_i + \sum_{j=1}^{mn-m-n+1} \epsilon_{ij} x_j, \quad 1 \leq i \leq 2r.$$

These expression can be replaced by

$$y_i = k_i + \sum_{j=1}^{mn-m-n+1} \epsilon_{ij} x_j + (-1)^i x_1, \quad 1 \le i \le 2r.$$

That is to say, the expressions for the eliminated unknowns $y_1, \ldots,$ y_{m+n-1} are not unique, which leads to a contradiction. Hence there does not exist a closed path in X whose vertices belong to Y.

2) Suppose that there does not exist a closed path of X whose vertices belong to $Y = \{ y_i : 1 \le i \le m+n-1 \}$. Now we proceed to prove that Y is a set of eliminated unknowns. If a row (or a column) of X contains only one element of Y, then the expression for this element may be obtained from a single equation in (10.3). For example, if the first row of X contains such an element x_{1j}, then

$$x_{1j} = a_1 - x_{11} - \cdots - x_{1,j-1} - x_{1,j+1} - \cdots - x_{1n}. \quad (10.6)$$

Denote the set of elements of Y with this property by Y_1. If, except for one element, all other elements of Y that occur in a fixed row (or column) of X belong to Y_1, then the expression for the exceptional element may be obtained from one equation of (10.3) and the expressions for certain elements of Y_1. For example, if the first row of X has such an exceptional element x_{1j}, then the expression for x_{1j} is obtained by substituting into (10.6) the right hand side of the (10.6) expressions for the elements of Y_1. The set of elements of Y with this property is denoted by Y_2. Continuing this process, if expressions for all elements of Y are obtained, then the assertion is true. In fact, since the expressions for the elements of Y are obtained using $m+n-1$ equations of (10.3), these expressions are unique. Hence Y is a set of eliminated unknowns.

If there is a $y_{s_1} \in Y$ for which an expression can not be obtained by the above process, then it is clear that another element y_{s_2} of Y with the same property can be found in the row where y_{s_1} is located. Associate this row with y_{s_1}. Similarly, an element y_{s_3} of Y with the same property can be found in the same column as y_{s_2}. Associate this column with y_{s_2}. Continuing this process, we obtain a sequence

$$y_{s_1}, y_{s_2}, y_{s_3}, \cdots,$$

with the associated rows and columns. Since there are only $m+n$ rows and columns, there will be a first y_{s_i} whose associated row or column is already associated with a previous y_{s_j}. It follows that

$$y_{s_i}\, y_{s_{j+1}} \cdots\, y_{s_{i-1}}\, y_{s_i}$$

is a closed path whose vertices belong to Y. This gives a contradiction, and the theorem is proved.   □

Let Y be a set with $m+n-1$ elements of X. Consider the system of equations

$$c_i + d_j = c_{ij}, \tag{10.7}$$

where $x_{ij} \in Y$. The system has $m+n-1$ equations and $m+n$ unknowns. Given any c_{i_0} (or d_{j_0}), if (10.7) has a unique solution, then Y is said to have <u>criterion numbers</u>.

<u>Lemma</u> 10.1. Suppose we have r lattice points in the coordinate plane such that there does not exist a closed path whose vertices are all among these lattice points. Through each lattice point, draw lines parallel to the x-axis and to the y-axis. Then there are at least $r+1$ distinct lines.

Proof. If $r = 1$, then we may draw two lines and so the lemma is true. Suppose that the lemma holds for $r = k$, where $k \geq 1$. Now we proceed to prove that the lemma holds also for $r = k+1$. Suppose there are at most $k+1$ distinct lines passing through the $k+1$ lattice points. There exists a line which passes through only one of these points, otherwise there exists a closed path whose vertices are all lattice points and this contradicts the assumption. Neglect this point and the corresponding line. Then there are at most k lines passing through the remaining k points. This leads to a contradiction with the inductive hypothesis. The lemma follows. □

Theorem 10.2. Y has criterion numbers if and only if there does not exist a closed path of X whose vertices belong to Y.

Proof. 1) Suppose that Y has criterion numbers and that there exists a closed path of X whose vertices belong to Y, say

$$x_{i_0 j_0} \; x_{i_0 j_1} \; x_{i_1 j_1} \cdots x_{i_{k-1} j_{k-1}} \; x_{i_{k-1} j_0} \; x_{i_0 j_0}.$$

Consider the system of equations

$$c_{i_0} + d_{j_0} = c_{i_0 j_0} \, ,$$
$$c_{i_0} + d_{j_1} = c_{i_0 j_1} \, , \qquad\qquad (10.8)$$
$$c_{i_1} + d_{j_1} = c_{i_1 j_1} \, ,$$
$$\vdots$$
$$c_{i_{k-1}} + d_{j_0} = c_{i_{k-1} j_0} \, ,$$

corresponding to these vertices. Since

$$c_{i_0} + d_{j_0} = (c_{i_0} + d_{j_1}) - (c_{i_1} + d_{j_1}) + \cdots$$

$$- (c_{i_{k-1}} + d_{j_{k-1}}) + (c_{i_{k-1}} + d_{j_0}),$$

there are at most $2k-1$ independent equations in (10.8), and so the number of independent equations in (10.7) is at most $m+n-2$. But there are $m+n$ unknowns in (10.7), and therefore it is impossible to determine the remaining unknowns in (10.7) uniquely when only one c_i (or d_j) is given. This leads to a contradiction and so there does not exist a closed path whose vertices belong to Y.

2) Suppose that there does not exist a closed path whose vertices belong to Y. Starting from any point $x_{i_0 j_0}$ of Y, d_{j_0} is determined uniquely by the equation

$$c_{i_0} + d_{j_0} = c_{i_0 j_0},$$

given the value of c_{i_0}. If

$$x_{i_0 j_1}, \ldots, x_{i_0 j_p}, \text{ and } x_{i_1 j_0}, \ldots, x_{i_q j_0},$$

are the other elements of Y in the i_0-th row and the j_0-th column of X, then c_{i_s} ($1 \leq s \leq q$) and d_{j_t} ($1 \leq t \leq p$) can be determined uniquely by the equations

$$c_{i_s} + d_{j_0} = c_{i_s j_0}, \quad 1 \leq s \leq q,$$

$$c_{i_0} + d_{j_t} = c_{i_0 j_t}, \quad 1 \leq t \leq p.$$

Consider the elements of Y in the i_s-th row or the j_t-th column of X, where $1 \leq s \leq q$ and $1 \leq t \leq p$. Then, corresponding to each element, a c_i (or d_j) is determined in the same way. Continuing this process, if Y has no criterion numbers, then one of the following two cases must hold.

a) For a given c_{i_k} and $x_{i_k j_l} \in Y$, suppose that d_{j_l} is already determined by the equation corresponding to $x_{i_{k_1} j_l}$ of Y. That is to say the equation

$$c_{i_k} + d_{j_l} = c_{i_k j_l}$$

is unnecessary. Then we can find $x_{i_k j_{l_1}}$ and $x_{i_{k_1} j_l}$ that were used to determine c_{i_k} and d_{j_l}, and then $x_{i_{k_1} j_{l_2}}$ and $x_{i_{k_2} j_{l_1}}$ that were used to determine $c_{i_{k_1}}$ and $d_{j_{l_1}}$, and so on, and finally $x_{i_{k_n} j_0}$ and $x_{i_0 j_{l_m}}$ that were used to determine $c_{i_{k_n}}$ and $d_{j_{l_m}}$. Hence we have a closed path whose vertices belong to Y, namely

$$x_{i_0 j_0} \ x_{i_0 j_{l_m}} \ \cdots \ x_{i_k j_l} \ \cdots \ x_{i_{k_n} j_0} \ x_{i_0 j_0} \ .$$

This gives a contradiction.

b) Suppose that the above process stops at a certain step, i.e. c_{i_0}, $\ldots$, c_{i_k} and d_{j_0}, $\ldots$, d_{j_l} are the criterion numbers that can be determined and there are c_i and d_j that can not be obtained. Let (i,j) be the coordinates of x_{ij}. Then through each point of X, we may draw lines parallel to the x-axis and the y-axis. It follows from Lemma 10.1 that there are at least $k+l+2$ distinct lines passing through the points of Y which are used to determine c_{i_0}, $\ldots$, c_{i_k}, d_{j_0}, $\ldots$, d_{j_l}, and that there exist at least $m+n-k-l-1$ distinct lines passing through the remaining points of Y. Since the above process has stopped, these two sets of lines have nothing in common. And so there are at least $m+n+1$ distinct lines passing through the points of Y. But there are only $m+n$ lines passing through all points of X, which gives a contradiction, and the theorem is proved. □

From Theorems 10.1 and 10.2, we have immediately the following.

Theorem 10.3. Y is a set of eliminated unknowns if and only if Y has criterion numbers.

§ 10.4 A criterion for optimality.

Let $y_1, \ldots, y_{m+n-1}$ be a set of positive eliminated unknowns with expressions given in (10.5). Substituting (10.5) in (10.2), we have

$$F(\vec{x}) = M + \sum_{j=1}^{mn-m-n+1} \lambda_j x_j.$$

The scalar λ_j is called the characteristic number of x_j corresponding to the eliminated variables $y_1, \ldots, y_{m+n-1}$.

Theorem 10.4. If $\lambda_j \geq 0$ ($1 \leq j \leq mn-m-n+1$), then the feasible solution is an optimal solution.

Proof. For the feasible solution

$$\vec{x}_0 : y_i = k_i \ (1 \leq i \leq m+n-1), \ x_j = 0 \ (1 \leq j \leq mn-m-n+1)$$

we have

$$F(\vec{x}_0) = M.$$

On the other hand, it follows from $\lambda_j \geq 0$ and $x_j \geq 0$ ($1 \leq j \leq mn-m-n+1$) that

$$F(\vec{x}) \geq M,$$

and thus $\vec{x}_0$ is an optimal solution. The theorem is proved. $\square$

§10.5 Characteristic numbers.

Let Y be a set of eliminated unknowns.

<u>Lemma</u> 10.2. For any remaining unknown $x_{i_0 j_0}$ there exists a unique closed path in X that passes through $x_{i_0 j_0}$ and whose remaining vertices are eliminated unknowns.

Proof. 1) Associate the i_0-th row with the unknown $x_{i_0 j_0}$. Consider the equation

$$\sum_{t=1}^{n} x_{i_0 t} = a_{i_0}.$$

Substituting the expressions (10.5) for the eliminated unknowns into any equation of (10.3), we always obtain an expression in which all terms vanish, and so there exists at least one eliminated unknown, $x_{i_0 j_1}$ say, among $x_{i_0 1}, \ldots, x_{i_0 j_0 - 1}, x_{i_0 j_0 + 1}, \ldots, x_{i_0 n}$ whose expression (10.5) has a non-zero coefficient for $x_{i_0 j_0}$. Associate the j_1-th column with the unknown $x_{i_0 j_1}$. Now consider the equation

$$\sum_{s=1}^{m} x_{s j_1} = b_{j_1}.$$

Using the same argument, there must be another eliminated unknown, $x_{i_1 j_1}$ say, among $x_{1 j_1}, \ldots, x_{i_0 - 1 j_1}, x_{i_0 + 1 j_1}, \ldots, x_{m j_1}$ whose expression (10.5) has a non-zero coefficient for $x_{i_0 j_0}$. Associate the i_1-th row with the unknown $x_{i_1 j_1}$. Continuing this process, we obtain a sequence of eliminated unknowns (except for $x_{i_0 j_0}$)

$$x_{i_0 j_0}, \ x_{i_0 j_1}, \ x_{i_1 j_1}, \ x_{i_1 j_2}, \ \ldots. \tag{10.9}$$

with their associated rows or columns. Since there are only $m+n-1$ eliminated unknowns there will be a first eliminated unknown, x_{pq} say, in this sequence for which one of the following occurs.

a) The eliminated unknown x_{pq} has associated column j_0 (thus x_{pq} is in fact $x_{i_s j_0}$ for some s). Then we have the closed path

$$x_{i_0 j_0} \ x_{i_0 j_1} \ \ldots \ x_{i_s j_0} \ x_{i_0 j_0} \tag{10.10}$$

whose vertices are all eliminated unknowns except for $x_{i_0 j_0}$.

b) The row or column associated with x_{pq} is already associated with a previous element in the sequence. Say $x_{pq} = x_{i_s j_s}$ and has associated row i_s, and the i_s-th row is already associated with $x_{i_t j_t}$ where $t < s$. We have the closed path

$$x_{i_s j_s} \ x_{i_t j_{t+1}} \ \ldots \ x_{i_{s-1} j_s} \ x_{i_s j_s}$$

whose vertices are eliminated unknowns. This contradicts with Theorem 10.1, and so we must be in situation a).

2) Suppose that there are two distinct closed paths Z and Z_1 passing through $x_{i_0 j_0}$, whose other vertices are eliminated unknowns. We will show that when drawing horizontal and vertical lines through the lattice points corresponding with the vertices of Z and Z_1, there are at least as many lattice points as there are lines. In view of Lemma 10.1 there exists a closed path whose vertices are eliminated unknowns in $Z \cup Z_1$, and thus in Y. This contradicts (Theorem 10.1) the fact that Y

is a set of eliminated unknowns.

Suppose Z contains r eliminated unknowns besides $x_{i_0 j_0}$. Since Z is a closed path, there are exactly $r+1$ horizontal and vertical lines passing through the $r+1$ lattice points corresponding with the vertices in Z, and hence there are exactly $r+1$ horizontal and vertical lines passing through the r lattice points corresponding with the vertices that are eliminated unknowns in Z. Using this as a start we form the set of all lattice points corresponding with the eliminated unknowns in $Z \cup Z_1$ as well as the corresponding set of horizontal and vertical lines by traversing through Z_1 starting at the first eliminated unknown after $x_{i_0 j_0}$. Each time an eliminated unknown is encountered that is not in Z, there will be at most one new line through the corresponding lattice point that has not yet been seen before. Now observe that the last vertex in Z_1 that is not in Z will not contribute a new line, and hence there are at least as many lattice points as lines. The lemma is proved. $\square$

Lemma 10.3. The remaining unknown $x_{i_0 j_0}$ does not appear in the expression (10.5) of any eliminated unknown besides those that are vertices in the closed path passing through $x_{i_0 j_0}$.

Proof. Suppose that the lemma does not hold. Then there is an eliminated unknown y_0 which is not a vertex of the closed path passing through $x_{i_0 j_0}$ and whose expression (10.5) has a non-zero coefficient for $x_{i_0 j_0}$. Similar to the proof of Lemma 10.2, we may prove that there is a closed path Z with vertex y_0 such that one of the following cases holds.

1) All vertices of Z are eliminated unknowns.

2) All vertices of Z except $x_{i_0 j_0}$ are eliminated unknowns.

Case 1) is contradicted by Theorem 10.1 and case 2) is contradicted by Lemma 10.2. This proves the lemma. $\square$

<u>Theorem</u> 10.4. Let $c_1, \ldots, c_m$ and $d_1, \ldots, d_n$ be a set of criterion numbers for the set of eliminated unknowns $\{y_1, y_2, \ldots, y_{m+n-1}\}$. Then the characteristic number $\lambda_{i_0 j_0}$ for the remaining unknown $x_{i_0 j_0}$ satisfies

$$\lambda_{i_0 j_0} = c_{i_0 j_0} - c_{i_0} - d_{j_0}.$$

Proof. It follows from Lemma 10.2 that there is a unique closed path (10.10) passing through $x_{i_0 j_0}$. After deleting the dependent equation

$$\sum_{i=1}^{m} x_{i j_0} = b_{j_0},$$

and using Lemma 10.3, we obtain the expression for $x_{i_0 j_1}$

$$x_{i_0 j_1} = -x_{i_0 j_0} + \cdots.$$

Substituting the expression (10.5) into the equation

$$\sum_{i=1}^{m} x_{i j_1} = b_{j_1},$$

we have an identity, and so from Lemma 10.3, we have

$$x_{i_1 j_1} = x_{i_0 j_0} + \cdots$$

Similarly, we may prove that the coefficients of $x_{i_0 j_0}$ in the expressions for $x_{i_1 j_2}, x_{i_2 j_2}, \ldots$ are alternately equal to -1 and $+1$. Therefore, by substituting (10.5) into (10.3), we obtain the expression for the characteristic number of $x_{i_0 j_0}$ as follows:

$$\lambda_{i_0 j_0} = c_{i_0 j_0} - c_{i_0 j_1} + c_{i_1 j_1} - \cdots - c_{i_s j_0} =$$

$$= c_{i_0 j_0} - (c_{i_0} + d_{j_1}) + (c_{i_1} + d_{j_1}) - \cdots - (c_{i_s} + d_{j_0})$$

$$= c_{i_0 j_0} - c_{i_0} - d_{j_0}.$$

The theorem is proved. □

Remark. As a corollary, it follows that ϵ_{ij} in the expression (10.5) is equal to 0, $+1$ or -1.

§ 10.6. Substitution.

If there is a remaining unknown $x_j = x_{i_0 j_0}$, say, with characteristic number $\lambda_j = \lambda_{i_0 j_0} < 0$, then the set of positive eliminated unknowns should be changed in the following way. Let (10.10) be the unique path of X passing through $x_{i_0 j_0}$ whose other vertices are eliminated unknowns. It follows from Lemma 10.3 that the coefficient of $x_{i_0 j_0}$ in the expression for an odd vertex of (10.9) is equal to -1, and equal to $+1$ for an even vertex, i.e. it takes on the value -1 at $x_{i_0 j_1}, x_{i_1 j_2} \ldots, x_{i_s j_0}$ and the value $+1$ at $x_{i_1 j_1}, x_{i_2 j_2}, \ldots,$ $x_{i_s j_s}$. And $x_{i_0 j_0}$ does not appear in the expressions of the other eliminated unknowns. Without loss of generality, we may suppose that the expression

$$x_{i_l j_{l+1}} = k - x_{i_0 j_0} + \cdots \tag{10.11}$$

has the smallest constant term among the expressions for the odd vertices of (10.10). Define $x_{i_l j_{l+1}}$ to be a remaining unknown and let $x_{i_0 j_0}$ be an eliminated unknown. The other eliminated unknowns are unchanged. Then we have a new set of eliminated unknowns. The expression of

$x_{i_0 j_0}$ may be derived directly from (10.11), namely

$$x_{i_0 j_0} = k - x_{i_l j_{l+1}} + \cdots . \qquad (10.12)$$

Substituting (10.12) into (10.5), we obtain the expressions for the other eliminated unknowns. The constant term in the expression of an even vertex in (10.10) is increased by the number k, while the constant term for an odd vertex is decreased by the number k. The expressions for all other eliminated unknowns are unchanged. Thus the new set of eliminated unknowns is still a set of positive eliminated unknowns. Substituting its corresponding feasible solution into (10.2), the value of $F(\vec{x})$ is decreased by the number

$$-k(c_{i_0 j_0} - c_{i_0 j_1} + \cdots - c_{i_s j_0}) = -k \lambda_j .$$

Let $\Sigma' a_i$ and $\Sigma' b_j$ denote the sums of certain elements of $a_1, \ldots, a_{m-1}$ and $b_1, \ldots, b_n$ respectively.

<u>Lemma</u> 10.4. The constant term in the expression of each eliminated unknown is of the form $\pm(\Sigma' a_i - \Sigma' b_j)$.

Proof. The expressions for a set of eliminated unknowns can be obtained from any $m+n-1$ equations of (10.3). Without loss of generality, we neglect the equation

$$\sum_{j=1}^{n} x_{mj} = a_m .$$

The expressions for the eliminated unknowns may be obtained by the process described in the proof of Theorem 10.1. The constant term in the expression for an element belonging to Y_1 is a_i or b_j, and the constant term in the expression for an element of Y_2 is of the form

$$a_i - \Sigma' b_j \quad \text{or} \quad b_j - \Sigma' a_i,$$

and so on. The lemma follows. □

The value of $F(\vec{x})$ would be decreased by a positive number

$$\left| \lambda_{i_0 j_0} (\Sigma' a_i - \Sigma' b_j) \right|$$

by a substitution of the above form if all terms of the form $\Sigma' a_i - \Sigma' b_j$ are not equal to zero. Since the number of feasible solutions is at most C_{m+n-1}^{mn} and the value of $F(\vec{x})$ is strictly decreased by such a substitution of a feasible solution, an optimal solution can always be obtained by a finite number of substitutions.

If there is a term $\Sigma' a_i - \Sigma' b_j$ that is equal to zero, then problem is said to be <u>degenerate</u>, and it may be solved by the following perturbation method. Let

$$\underset{\Sigma' a_i - \Sigma' b_j \neq 0}{\text{MINIMUM}} \left| \Sigma' a_i - \Sigma' b_j \right| = (n+1)\delta.$$

Let

$$b_j' = b_j + \delta, \quad 1 \leq j \leq n,$$

$$a_i' = a_i, \quad 1 \leq i \leq m-1, \quad \text{and} \quad a_m' = a_m + n\delta.$$

Then

$$\sum_{i=1}^{m} a_i' = \sum_{j=1}^{n} b_j'.$$

and

$$\Sigma' \, a_i' - \Sigma' \, b_j' = \Sigma' \, a_i - \Sigma' \, (b_j + \delta)$$

$$= \Sigma' \, a_i - \Sigma' \, b_j - k\delta \neq 0,$$

where $0 \leq k < n+1$.

Consider the new problem: Find x_{ij}, $1 \leq i \leq m$, $1 \leq j \leq n$, such that

$$G(\vec{x}) = \sum_{j=1}^{n} c_{ij} x_{ij} \qquad (10.13)$$

attains its minimum, and where the x_{ij}'s satisfy

$$\sum_{j=1}^{n} x_{ij} = a_i', \ 1 \leq i \leq m,$$

$$\sum_{i=1}^{m} x_{ij} = b_j', \ 1 \leq j \leq n$$

and

$$x_{ij} \geq 0, \ 1 \leq i \leq m, \ 1 \leq j \leq n.$$

This new problem is non-degenerate. Substituting the expressions for a set of eliminated unknowns into (10.13) and the corresponding expressions for the old problem into (10.2), we see that the characteristic numbers of the remaining unknowns are the same. Hence an optimal solution of the original problem can be induced from an optimal solution for the new problem.

Remark. The perturbation method is usually not needed in practice.

§ 10.7 Linear programming.

Let A be a $m \times n$ matrix with real coefficients. Let $\vec{c}$ and $\vec{b}$ be row vectors of dimensions n and m respectively.

Problem. Determine a n-dimensional vector $\vec{x}$ for which the objective function

$$\vec{c}\,\vec{x}^{T} \tag{10.14}$$

attains its minimum, where $\vec{x}$ satisfies the constraints

$$A\,\vec{x}^{T} = \vec{b}^{T} \tag{10.15}$$

and

$$\vec{x} \geq \vec{0}. \tag{10.16}$$

In general, (10.15) and (10.16) determine a polyhedron and (10.14) is a linear function, and therefore, if $\vec{c}\,\vec{x}^{T}$ has a minimum subject to (10.15) and (10.16), it can be attained at the vertices of the polyhedron. If (10.15) can be written as

$$\vec{y}^{T} = \vec{d}^{T} + B\,\vec{z}^{T}, \tag{10.17}$$

where $\vec{y} = (y_1, \dots, y_m)$, $\vec{z} = (z_1, \dots, z_{n-m})$, B is a $m \times (n-m)$ matrix and $\vec{d} = (d_1, \dots, d_m)$ is a vector with non-negative components, then $\vec{y}$ is called a set of positive eliminated unknowns and $\vec{z}$ a set of remaining unknowns, and thus we have a solution of (10.15) and (10.16),

$$\vec{y} = \vec{d}, \; \vec{z} = \vec{0}. \tag{10.18}$$

It is called a feasible solution. A feasible solution $\vec{x}$ is called an optimal solution if $\vec{c}\,\vec{x}$ is minimal. Substituting (10.17) into (10.14), we have

$$\vec{c}\,\vec{x}^T = M + \Sigma\,\lambda_j\,z_j.$$

If all λ_j's are non-negative, then (10.18) is an optimal solution. Otherwise, if there is a $\lambda_j < 0$, then, similar to the transportation problem, one of the positive eliminated unknowns should be replaced by another one so that the value of the objective function $\vec{c}\,\vec{x}^T$ is decreased. If the problem is non-degenerate, an optimal solution is obtained in finitely many substitutions. Otherwise, it can be treated also by the perturbation method. This is the well-known simplex method for linear programming.

References.

Dantzig G. B. Linear Programming and Extensions, Princeton Univ. Press, N.J., 1963.

Hitchcock, F.L. Distribution of a Product from several Sources to numerous Localities, J. of Math. Phy, 1941, 224-230.

Kantorowitch L. V. Mathematical Methods in the Organization and Planning of Production, Pub. House of the Leningrad State University, 1939.

Wan Zhe Xian and Wang Yuan. Mathematical Methods in Transportation Problem, Science Press, Beijing, 1959.

Yu Ming-I, Wan Zhe Xian and Wang Yuan etc. (edited). The Theory and Application of Linear Programming, People's Education Press, Beijing, 1959.

Editor's note: Besides the smallest element method, commonly used rules to find an initial transportation schedule are the North West Corner Rule, Vogel's approximation and Russell's approximation. The criterion numbers are usually refered to as the dual variables and substitution is often called pivoting. The eliminated unknowns are often called the basic variables, the remaining unknowns are the non-basic variables, and a feasible solution is called a basic feasible solution. The characteristic numbers are the reduced costs. Among the many textbook treatments of linear programming and the transportation problem we mention the following.

Bazaraa, M.S. and J.J. Jarvis. Linear Programming and Network Flows. John Wiley & Sons, Inc., 1977.

Hillier F.S., and G.J. Lieberman. Introduction to Operations Research. 4-th ed., Holden-Day Inc., 1986.

CHAPTER 11

THE POSTMAN PROBLEM

§ 11.1 Introduction.

Every day a postman will start from the post office and traverse through certain streets to deliver and collect letters and then return to the post office. A natural question to ask is whether there is a rule by which we can find a route that includes all streets he has to pass through so that the length of the route is a minimum. The routing of the garbage trucks in large cities is a similar problem.

The route traversed by the postman may be regarded as an undirected connected graph, where streetcorners are nodes and streets are arcs. The number of arcs incident to a node is called <u>the degree of the node</u>. A node with odd degree is called <u>an odd node</u>, otherwise it is called <u>an even node</u>. For example, A, B, D and F are odd nodes and C, E, G, and H are even nodes for the graph in Figure 11.1.

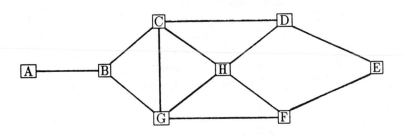

Figure. 11.1

202

A famous problem considered by Euler is whether a given graph has a walk through the graph that traverses each arc exactly once and returns to the initial point. If such a walk exists, it is said that the graph has an <u>Euler</u> <u>path</u> or Euler tour. Clearly, if the route to be traversed by the postman has an Euler path, then such a path is a route with least length. Euler proved in 1736 that a graph has an Euler path if and only if each node of the graph is even.

Hence if all nodes of a graph are even, then an optimum walk is obtained. Otherwise, for any graph with odd nodes it is possible to select a set of arcs that will be duplicated such that the resulting graph does satisfy the condition for having an Euler path, for example duplicating the entire set of arcs will work. However, there are many ways to select sets of arcs for duplication so that the resulting graph has an Euler tour. For example, duplicating the arcs AB, DE, and EF or the arcs AB, DH, FH results in the graphs of Figures 11.2 and 11.3, and both have Euler tours. The length of the arcs are shown in Figure 11.3. The total length of the duplicated arcs is 5 in Figure 11.2 and 3 in Figure 11.3.

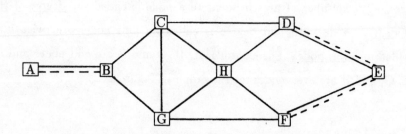

Figure 11.2

Hence, if a graph has odd nodes, the problem becomes how to find a set of arcs for duplication in order to yield an Euler path such that the total length of the duplicated arcs is a minimum. This is the problem faced by a postman in choosing a minimum length route to bring him back to the post-office.

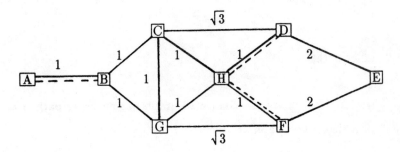

Figure 11.3

This problem is called the Postman Problem (or the Chinese Postman Problem in foreign literature) and it was first defined by Guan Mei Gu in 1958, who was inspired by the graphical method for the transportation problem. He proved that the resulting graph has an Euler path with minimum total length if and only if the total length of the duplicated arcs in each cycle of the graph does not exceed half the length of the cycle (a cycle is a closed path that does not self-intersect, for example $DEFH$ in Figure 11.1).

If the total length of the duplicated arcs in a cycle C of the graph is greater than half the length of C, then we may duplicate instead the other arcs of C. The resulting graph has also an Euler path but the total length of the arcs has decreased. Since for any given graph, there are only a finite number of ways to select a set of arcs for duplication, we may obtain, by finitely many substitutions, a graph having an Euler path such that the total length of the arcs is a minimum.

However, the situation is similar to the graphical method for the transportation problem. If the number of cycles in a graph is comparatively large, the method is certainly inconvenient, and therefore a better computational method will be needed, but we will not develop one here.

§ 11.2 Euler paths.

<u>Theorem</u> 11.1. An undirected connected graph has an Euler path if and only if all nodes of the graph are even.

Proof. 1) First, we will prove that if all nodes are even, then the graph has an Euler path. Let n denote the number of nodes of the graph. If $n = 2$ and there are $2m$ arcs incident to the two nodes, A and B say, then we may traverse from A to B using m arcs and return from B to A using the remaining m arcs, and therefore the graph has an Euler path. Now suppose that $n > 2$ and that the assertion holds for graphs with less than n nodes. Now we proceed to prove that the assertion holds also for graphs with n nodes. For a given node A, suppose that the arcs incident with A are $AB_1, \ldots, AB_{2m}$. Neglect the $2m$ arcs $AB_1, \ldots, AB_{2m}$, delete the node A and add m arcs $B_1B_2, \ldots, B_{2m-1}B_{2m}$. The resulting graph has $n-1$ even nodes, and therefore by the induction hypothesis, this graph has an Euler tour. Since the route along the arcs $B_1B_2, B_3B_4, \ldots, B_{2m-1}B_{2m}$ can be changed to $B_1AB_2, B_3AB_4, \ldots, B_{2m-1}AB_{2m}$ respectively, the original graph has also an Euler path, and thus the assertion follows by induction.

2) Next, we will prove that a graph with odd nodes does not have an Euler tour. Suppose that A is an odd node and there are $2m+1$ arcs incident with it. Let us start from A. It is evident that there does not exist a walk that would traverse each arc exactly once and return to A. Hence the theorem is proved. □

§ 11.3 A necessary and sufficient criterion for an optimum solution.

Problem: Given any undirected connected graph G, find a walk
that starts at a given node, traverses every arc and returns to the initial
node so that the total length of the route is a minimum.

If all nodes of G are even, then, by Theorem 11.1, G has an
Euler path which satisfies the requirements of our problem. Otherwise, a
set of arcs will be selected for duplication such that the resulting graph
F has only even nodes. The graph F is called a feasible solution of G.
A feasible solution with minimum total arc length is called an optimum
solution of G. If the total length of the duplicated arcs in each cycle of
F does not exceed half the length of the cycle, then F is called a
normal solution.

Theorem 11.2. A feasible solution of G is optimum if and only if it is
normal.

To prove Theorem 11.2, we will need

Lemma 11.1. The number of odd nodes of any undirected connected
graph H is even.

Proof. If H has s odd nodes A_i ($1 \leq i \leq s$) and t even nodes B_j
($1 \leq j \leq t$), and the number of arcs incident to A_i and B_j are a_i
and b_j respectively, then

$$\sum_{i=1}^{s} a_i + \sum_{j=1}^{t} b_j$$

equals two times the number of arcs of H. Hence

$$\sum_{i=1}^{s} a_i + \sum_{j=1}^{t} b_j \equiv 0 \pmod 2,$$

and hence s is even. The lemma is proved. $\qquad\qquad\qquad\square$

Proof of Theorem 11.2. 1) Suppose that a feasible solution F_1 of G is not normal. Then G has a cycle C for which the total length ℓ of arcs on C that were duplicated to obtain F_1 exceeds half the length L of C, i.e.

$$\ell > \tfrac{1}{2} L.$$

Let F_2 be the graph which differs from F_1 only in the sense that the arcs of C duplicated according to F_1 are not duplicated in F_2 and the other arcs of C are duplicated in F_2. Then F_2 is also a feasible solution of G but the total length of the arcs of F_2 is less than that of F_1, and their difference is equal to

$$\ell - (L - \ell) = 2\ell - L.$$

Therefore F_1 is not an optimum solution for G.

2) The above process of improving F_1 to F_2 is called a substitution of a feasible solution. Since the total number of feasible solutions is finite and the total length of a feasible solution is strictly decreased by such a substitution, we may always obtain a normal solution by a finite number of substitutions. If G has only one normal solution, it is evidently the optimum solution. Otherwise, if G has at least two normal solutions, it suffices to prove that the total length of all arcs for any two normal solutions, F and F' say, are equal. By Lemma 11.1, we may suppose that G has $2n$ odd nodes and that the duplicated arcs form n arc-disjoint paths by which the odd nodes are

pairwise connected.

Suppose that $n = 1$, i.e. G has only two odd nodes, A and B say. Then we have a closed path Z consisting of the arcs duplicated according to F from A to B and those duplicated according to F' from B to A. The path Z contains arcs that are common to F and F' and the remaining arcs of Z form a set of arc-disjoint cycles. In each cycle, the total length of the arcs duplicated according to F and the total length of the arcs duplicated according to F' are the same, namely half the length of the cycle. Hence F and F' have the same total length.

Now suppose that G has $2n$ odd nodes, where $n > 1$, and that the assertion holds for any undirected connected graph whose number of odd nodes is less than $2n$. Using the n arc-disjoint paths of duplicated arcs for F and for F', we do the following. Starting from an odd node A_1 of G, traverse along the arcs duplicated according to F to an odd node B_1 of G, then from B_1 traverse along the arcs duplicated according to F' to an odd node A_2 of G, and then traverse along the arcs duplicated according to F from A_2 to an odd node B_2 of G, and so on. We obtain a sequence of odd nodes of G, namely

$$A_1, B_1, A_2, B_2, \ldots .$$

Since G has only $2n$ odd nodes, there exist two nodes A_s and A_t such that

$$A_s = A_t, A_i \neq A_j \quad (s \leq i < j < t \text{ or } s < i < j \leq t).$$

Hence we have a closed path

$$Z: A_s B_s \ldots A_{t-1} B_{t-1} A_t \, (= A_s)$$

which contains arcs that are common to F and F' and whose other arcs form a set of arc-disjoint cycles. Similar to the case $n = 1$, we may prove that the total length of arcs of F and the total length of arcs of F' in Z are the same. Suppose that G' is obtained from G by duplicating the paths A_sB_s, $A_{s+1}B_{s+1}$, $A_{t-1}B_{t-1}$. Then G' is an undirected connected graph, where the odd nodes A_s, B_s, ..., A_{t-1}, B_{t-1} are changed into even nodes and the other nodes of G remain unchanged. Thus G' is a graph with $2n - 2(t - s)$ odd nodes. Let F^* be the graph obtained from F' by deleting the arcs on Z duplicated according to F', and duplicating the arcs on Z duplicated according to F. Then F^* is also a feasible solution of G with the same length as F'. Since F and F^* are normal solutions of G', it follows by induction that F and F^* have the same length, and therefore F and F' have the same length. The assertion follows by induction, and so the theorem is proved. $\square$

References.

Guan Mei Gu. Graphic Programming using Odd and Even Points. Acta Math. Sinica, 10, 3, 1960, 263-266. (also: Chinese Math 1, 1962, 273-277).

Editor's note: An algorithm for the Postman Problem that runs in polynomial time was developed by J. Edmonds in 1965 and is in essence an application of Edmonds' polynomial time algorithm for the Weighted Matching Problem. For a description of these results, see

Lawler, E.L. Combinatorial Optimization: Networks and Matroids. Holt, Rinehart and Winston, 1976.

For some recent results on the Postman Problem, see

Lovász, L. and M.D. Plummer. <u>Matching Theory</u>. Annals of Discrete Mathematics, 29, North Holland, 1986.